Industrial Organic Chemistry

Laboratory Manual

John M. Kokosa, Ph.D.

Kettering University

KENDALL/HUNT PUBLISHING COMPANY
4050 Westmark Drive   Dubuque, Iowa 52002

Copyright © 2002 by John M. Kokosa

ISBN 978-0-7872-9602-5

Printed in the United States of America
10 9 8 7 6 5

Table of Contents

Dedication

This text is dedicated to my parents, wife, children and all my students and colleagues who had the patience to allow me to bring this work to completion.

Acknowledgments

I would especially like to acknowledge my secretary, Ms Lenore Myers, who spent countless hours on countless revisions of this manual. I would also like to thank Mr. John Kirby for fine tuning many of the experiments and my colleagues and students who made positive suggestions for changes and refinements for almost every page of the text.

SAFETY CONTRACT

The purpose of this contract is to define acceptable behavior in the academic laboratory and to make the student aware of his/her responsibility for laboratory safety.

I, _____, will:
(print name)

✔ follow all instructions given by the laboratory instructor.

✔ protect eyes, face, hands and body when in the laboratory.

✔ know the location of safety equipment.

✔ conduct myself in a responsible manner at all times.

I have read and agree to follow the safety rules and regulations set forth above and on the safety sheet for _____ Laboratory.
(print Lab Course & Section Number)

I will closely follow the written and oral instructions given by the laboratory instructor.

Date:_____ Signed:_____

SAFETY IN THE CHEMICAL LABORATORY

Safety considerations play an important part in the planning and execution of all laboratory experiments. A disregard for safety procedures can cause delays, endanger yourself and others, and can lead to unreliable data and incorrect interpretations.

The laboratory rules have been established to insure that you, other students in the lab and your instructor have a safe and productive laboratory experience.

Safety in the laboratory consists of following a set of safety rules, following correct procedures as outlined by your instructor, using the CORRECT equipment for each experiment and knowing what hazards are involved with each material used. It is important that you treat all chemicals in the laboratory as potentially hazardous, and report any spills to your laboratory instructor, who will determine the proper cleanup methods.

The following **Safety Rules** are followed in **all** chemistry laboratories:

1. **Chemical splash goggles must be in place before entering the laboratory and must be worn at all times while in the laboratory, even if you wear contacts or eyeglasses.**

2. Confine hair that reaches the shoulders. Hair is extremely flammable, and can react with a variety of compounds found in the laboratory.

3. **DO NOT** eat, drink or taste anything in the laboratory. **DO NOT** bring any food or beverages into the laboratory. No chemicals are ever to be taken from the laboratory.

4. Do only the experiment assigned and in the manner **described by your instructor**. Unauthorized experiments are prohibited. Horse play and practical jokes in the laboratory are dangerous and absolutely prohibited.

5. Correct dress for the laboratory includes long pants (or long skirts or lab aprons) covering the knees; shirts with sleeves, and footwear that completely covers the foot. Older clothing should be worn and laboratory aprons are always available. No shorts, sleeveless shirts, halter tops, sandals, etc. are permitted.

6. Avoid inhaling chemical fumes and **never** pipet by mouth! Do not sit or stand on the laboratory benches as they are often contaminated.

7. **Do not** allow laboratory chemicals to come into contact with your skin. If this occurs, the chemical should be removed **immediately** in a manner prescribed by your instructor. **All** chemical spills (no matter how small) that occur during the laboratory period should be reported to your instructor and then cleaned up in the manner prescribed by the instructor. The instructor is to be notified immediately after any incident, no matter how trivial it may seem.

8.   **DO NOT** attempt to clean up blood that results from a cut, nor render first aid. If you are cut immediately apply a compress (clean paper towel) and inform the laboratory instructor. **DO NOT** attempt to pick up glass with your hands. Broken glass is picked up either by sweeping into a dust pan or using tongs to pick up larger pieces. **All broken glass is placed into the broken glass boxes. No material is to be retrieved from the broken glass boxes !**

9.   Use the waste containers provided in the laboratory. Make sure you understand the **proper** disposal procedures for the chemicals you are using. **DO NOT** use the sinks to discard matches, filter paper or any solid wastes.

10. Never point the open end of a test tube towards yourself or others.

11. **DO NOT** return chemicals to their original container. Treat excess material as waste and dispose of it accordingly.

12. Use common sense when working with chemicals. If you are unsure about procedures, **ask your instructor**. No student is permitted to work in the laboratory alone.

13. **Always** wash your hands before **leaving** the laboratory area. Soap dispensers are located throughout the laboratory.

14. If a student is pregnant, or becomes pregnant during the term, the instructor **must be notified immediately** so that, if necessary, alternative arrangements can be made to complete the laboratory requirements.

15. Make sure you know where the following safety equipment is located in your lab:

Fire extinguishers	Shower
Emergency exits	Emergency telephone
Material Safety Data Sheets	Sinks for washing glassware
Coat rack/book bag racks	Broken glass container
Waste receptacles	Soap dispensers
Paper towel dispensers	Lab spill clean up kit
De-ionized water faucets	Hoods
Hazardous Waste Containers	Ice Dispenser
Eye wash	

16.   In case of a fire alarm, the following procedures will be followed:

    a.   Turn off all equipment.
    b.   Exit through hall lab entrance. Use the stairs to exit out of the building.
    c.   If the main route is blocked, follow the instructor's direction for alternate routes.

17.   In case of a tornado alarm, the following procedures will be implemented:

    a.   Turn off all equipment
    b.   Exit through hall lab entrance, use the stairs proceed to the designated tornado shelter area, as instructed.

Material Safety Data Sheets

Material Safety Data Sheets (MSDS) were mandated by the Occupational Safety and Health Administration as part of a program designed to provide information on hazardous chemicals to employees who work with them as part of the job. The MSDS are located in the laboratories in order to help faculty/staff/students understand the hazards of the chemicals they are working with, the types of personnel protective equipment that might be used when working with these chemicals, and what to do in case of a spill of these chemicals, and how to safely dispose of these chemicals.

The MSDS for all the chemicals used in the chemistry laboratories are available in the Chemistry Instrumentation Room. There is a binder which will contain the MSDS for all of the chemicals you will be using during the course of this laboratory term. These are available during the regular laboratory periods. In addition, the master MSDS File is located in the office of Environmental Health and Safety.

Although there is no standard format for the MSDS, the following information is usually found on all MSDS:

General Information:

Name and address of the manufacturer/distributor/importer. Emergency telephone number which must be answered 24 hours a day if the manufacturer claims a trade secret for the material.

Name of the product as it appears on the label.

Formula CAS Number. This is a number given to each unique chemical by Chemical Abstract Services. The chemical structure, physical and biological data, and route of manufacture can always be obtained even if the MSDS is vague on these points.

Precautionary/First Aid information in case of accidental exposure, injury or spill.

Hazardous Ingredients:

List of the hazardous ingredients in the material. There may be only one in the case of reagent chemicals. All chemicals that are greater than 1% by mass of the total mass must be listed. If the chemical is a carcinogen, it must be listed if it is greater than 0.1% by mass of the total product mass.

Physical Data:

Physical data for the material would include (if applicable) the boiling point, specific gravity, vapor pressure, vapor density, percent volatile (by volume), evaporation rate, solubility in water appearance and odor, physical state, pH, and any other physical data that the manufacturer finds significant.

Fire and Explosion Hazard Data:

This section includes the classification, flash point, lower explosion (flammability) limit (LEL,LFL), upper explosion (flammability) limit (UEL,UFL), extinguishing media, autoignition temperature, combustion products, special fire-fighting procedures and any other information a fire fighter may need.

Health Hazard Data:

This section includes information on how the material may get into the body, the effects of overexposure, the exposure limits, toxicity in terms of LD_{50}, emergency and first aid procedures and other health information.

Reactivity Data:

Information about how this material interacts with other materials and the environment are found in this section. Included is whether or not the material is stable under normal working conditions, incompatibilities with other materials, hazardous decomposition products, whether or not the material undergoes hazardous polymerizations and storage and use conditions to avoid.

Spill or Leak Procedures:

This section covers the steps to be taken in case of a spill or release into the environment. Often included in this section is what type of absorbents to use, type of respiratory protection required, and statutory requirements under the Superfund regulations, and disposal information.

Safe Handling and Protection Information:

Types of personal protective equipment (PPE) that should be used is detailed in this section along with the types of ventilation that need to be provided.

Special Precautions:

Other information on the handling of this material, special storage considerations, and other types of warning are found in this section.

Definition of Terms used on MSDS:

LD_{50} Lethal dose at which 50% of the test animals died when given the chemical either orally or applied dermally

LC_{50} Lethal air concentration at which 50% of the test animals died when breathing chemical gas or vapor (in ppm) or mist, dust or fumes (in mg/l)

Highly toxic LD_{50} (oral) <= 50 mg/kg body weight
LD_{50} (dermal) <= 50 mg/kg body weight
LC_{50} (inhalation) <= 200 ppm gas or vapor
             2 mg/l dust or mist

Toxic LD_{50} (oral) >50<500 mg/kg body weight
LD_{50} (dermal) >50<500 mg/kg body weight
LC_{50} (inhalation) >200<2000 ppm gas or vapor
             >2<20 mg/l dust or mist

Corrosive   A chemical that causes visible destruction of, or irreversible alterations in, living tissue by chemical action at the site of contact.

Irritant   A chemical, which is not corrosive, but which causes a reversible inflammatory effect on living tissue by chemical action at the site of contact.

Sensitize   A chemical that causes a substantial proportion of exposed people or animals to develop an allergic reaction in normal tissue after repeated exposure to the chemical.

Flammable   Having a flash point of less than 100°F (OSHA) or 140°F (DOT).

Flash Point   Temperature at which there is sufficient vapor to ignite when a spark is introduced.

Oxidizer   A chemical that has oxygen as part of its formula or can promote a reaction (examples would include KNO_3, F_2, Cl_2, O_2).

GETTING MSDS SHEETS VIA THE INTERNET

Log into the University System.

1. You can search through a standard search engines like Yahoo or Google for "MSDS" or "Materials Safety Data Sheets."

2. Try the University of Vermont web page (siri.uvm.edu)

3. The Vermont page has dirrect links to many university and company MSDS archives at siri.uvm.edu/msds/links.htm

4. The vermont page also has links to EPA, NIOSH and other health and safety links

5. Particularly useful links for msds sheets include the University of Utah (gopher://atlas.chem.utah.edu/11/MSDS) and Cornell University (msds.pdc.cornell.edu/ISSEARCH/MSDSsrch.HTM)

6. You will be able to find MSDS sheets for chemicals such as sulfuric acid and benzene, but probably not for trade names for products such as WU Etch, which must be provided directly from the manufacturer.

COURSE OBJECTIVES FOR
Industrial Organic Chemistry

The students should be able to do the following for simple alkanes, alkenes, aromatics, alcohols, phenols, ethers, halogenated hydrocarbons, aldehydes, ketones, amines, acids, amides and esters:

Lecture Course Objectives

1. Drawing structural formulas. Students should be able to draw and recognize compounds written using the following structural representations:

    a) expanded structural formulas

        example:

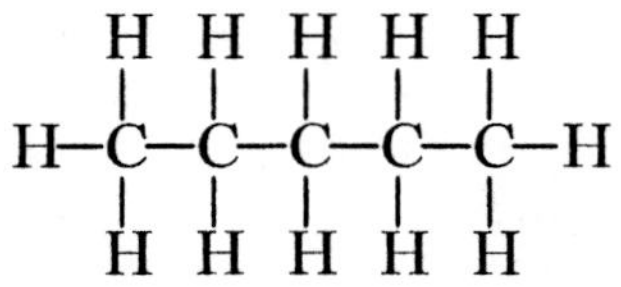

    b) condensed structural formulas

        examples: $CH_3-CH_2-CH_2-CH_2-CH_3$ or $CH_3-(CH_2)_3-CH_3$

    c) Skeletal formulas

        (pentane)    (2,2,4-trimethylpentane) 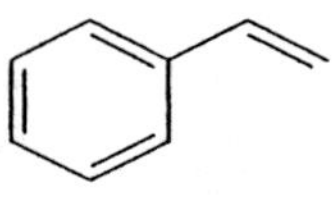   (styrene)

2. Identify the hybridization for atoms.

3. Identify sigma and pi bonds.

4. Know the bond angles present for sp, sp^2 and sp^3 hybridized systems.

5. Indicate bond polarity.

6. Indicate how polarity and general structure influence the following?

    a) boiling point

    b) vapor pressure

c)   solubility (miscibility)

7.   IR Spectra: Be able to do a first order interpretation of IR spectra for molecules containing the common functional groups.

8.   Give typical industrial/commercial uses for the major classes of organic compounds.

9.   Write reactions to illustrate:

a)   free radical halogenation of alkanes
        (Be able to draw the mechanism for this reaction).

b)   electrophilic addition to double bonds
        (Be able to draw the mechanism for this reaction).

c)   electrophilic aromatic substitution
        (Be able to draw the mechanism for this reaction).

d)   substitution of alcohols and alkyl halides

e)   elimination of alcohols and alkyl halides

f)   esterification

g)   hydrolysis

10.   Be able to draw structural formulas and know the common and IUPAC names of the following organic compounds.

a)   **alkanes**
      butane (n-butane)
      2-methylpropane (isobutane)
      2,2,4-trimethylpentane (isoctane)

b)   **alkenes**
      ethene (ethylene)
      propene (propylene)
      2-methylpropene (isobutylene)
      2-methyl-1,3-butadiene (isoprene)

c)   **alkynes**
      ethyne (acetylene)

d)   **aromatics**
      toluene
      (o, m, p) xylenes
      styrene

e) **alcohols**
methanol (methyl alcohol)
ethanol (ethyl alcohol)
2-propanol (isopropyl alcohol)
2-propen-1-ol (allyl alcohol)
phenylmethanol (benzyl alcohol)

f) **halogenated hydrocarbons**
chloromethane (methyl chloride)
dichloromethane (methylene chloride)
trichloromethane (chloroform)
tetrachloromethane (carbon tetrachloride)
2-chloropropane (isopropyl chloride)
chloromethylbenzene (benzyl chloride)
2-chloro-2-methylpropane   (t-butyl chloride)
chloroethene (vinyl chloride)
dichlorodifluoromethane (freon 12)
tetrachloroethene (tetrachloroethylene)

g) **phenols**
phenol
4-chlorophenol (p-chlorophenol)

h) **amines**
methylamine
ethylamine
triethylamine
aniline
pyridine

i) **aldehydes**
methanal (formaldehyde)
ethanal (acetaldehyde)
benzaldehyde

j) **ketones**
propanone (acetone)
butanone (MEK, methyl ethyl ketone)

k) **ethers**
diethyl ether
THF (tetrahydrofuran) (optional)
1,4-dioxane (optional)

l) **acids**
methanoic acid (formic acid)
ethanoic acid (acetic acid)
benzoic acid
p-toluenesulfonic acid

m) **acid anhydrides**
acetic anhydride
phthalic anhydride (optional)

n) **amides**
formamide (optional)
acetamide (optional)

o) **esters**
formates
acetates
benzoates

11.   Be able to name cyclic and double bond molecules according to the following conventions:

a)   To be able to label alkenes and cycloalkanes as *cis* or *trans*, for example:

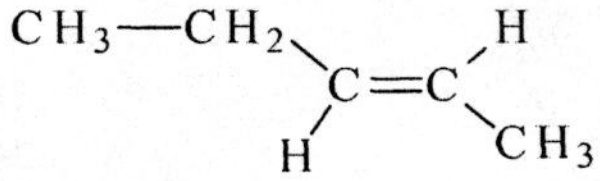

$$CH_3—CH_2 \diagdown \atop H \diagup C=C \diagup H \atop \diagdown CH_3$$

trans-2-pentene 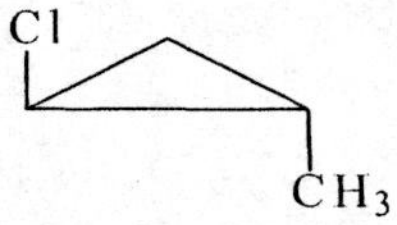   trans-1-chloro-2-methylcyclopropane

b)   To be able to label alkenes as E or Z, for example:

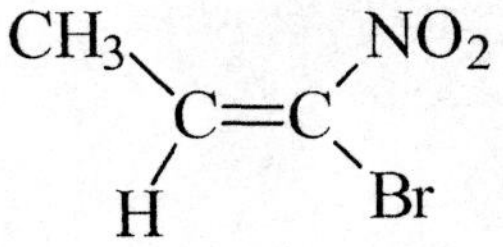

$$CH_3 \diagdown \atop H \diagup C=C \diagup NO_2 \atop \diagdown Br$$

$$CH_3—O \diagdown \atop H_2N \diagup C=C \diagup F \atop \diagdown CH_2—CH_3$$

E-1-bromo-1-nitropropene 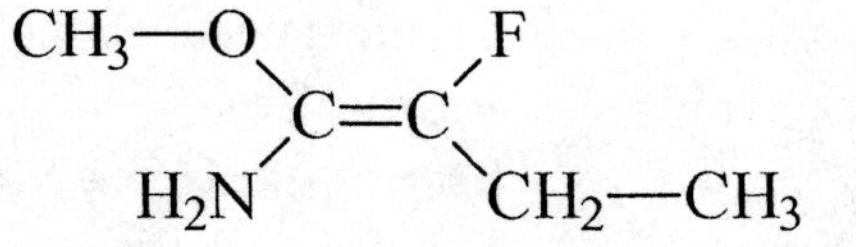   E-1-amino-2-fluoro-1-methoxy-1-butene

12.   The basic principles of polymer chemistry. The student should be able to:

a)   Distinguish between addition and condensation polymers based on their mechanism of formation and their structures.

b)   Name and draw the structural formulas of simple addition and condensation polymers. Examples:

polyethylene   polypropylene   poly(vinyl chloride)
polystyrene   polytetrafluoroethylene   nylon 6,6
poly(ethylene terephthalate)

Laboratory Course Objectives

1.   Be able to predict solubilities of hydrocarbons, alcohols, phenols, amines, aldehydes, ketones and carboxylic acids in water, acid, base and typical organic solvents, such as ethanol and hexane.

2.   Know the following visual tests to identify common functional groups:

Alkanes: Bromine Test (Br_2/CH_2Cl_2/hv or Δ)
Alkenes & Alkynes: Bromine Test (Br_2/CH_2Cl_2/25°C/Dark)
                          Baeyer Test ($KMnO_4$)
                          H_2SO_4
Aromatics: Sulfonation (H_2SO_4 / Δ)

Alcohols:	Lucas Test ($ZnCl_2$ / HCl)
	Chromic Acid Test
Phenols:	Litmus Test
	$FeCl_3$ Test
	Br_2 / H_2O
	Solubility in 5% NaOH
Ketones:	DNP Test
Aldehydes:	DNP Test
	Silver Mirror/Tollens' Test
Amines:	Litmus Test
	Solubility in 5% HCl
Hydrocarbons:	Flame Test for Unsaturation
Carboxylic Acids	Litmus Test
	Solubility in 5% $NaHCO_3$
All Families:	Solubility Tests

3. Be able to predict the relative Retention Factors (R_f values) for chemicals with varying polarities when using thin layer chromatography (TLC).

4. Be able to predict the relative retention times for chemicals with varying boiling points when using gas chromatography (GC).

5. Be able to identify simple polymers using a combination of physical-chemical properties, infrared spectroscopy (IR) and differential thermal calorimetry (DSC) experimental data.

6. Be able to identify simple, pure chemicals using a combination of boiling point, melting point, physical-chemical properties and IR experimental data.

Basic Rules of Nomenclature for Alkanes, Cycloalkanes, Alkenes, Alkynes and Aromatics

Summary

1. For an open chain alkane, find the longest continuous chain and number from the end which gives the lowest numbers to the substituents.

2. For an alkene or alkyne, the double or triple bond takes precedence in numbering over substituents. Numbering must proceed **through** the multiple bond. If both a double and triple bond are present (an enyne), number from the end nearer the first multiple bond. If numbering is the same in both directions, the tripple bond takes precedence. Numbers are placed either **directly** in front of the molecular fragment name they pertain to or directly in front of the suffix. The suffix for an alkene is **ene**, for an alkyne **yne**.

3. Cyclic alkanes, -alkenes, and -alkynes use the prefix cyclo. One of the double bond carbons of a cyclic unsaturated system is carbon #1 and numbering must proceed **through** the multiple bond. Otherwise, numbering is as in alkanes.

4. Aromatic systems are named as their parent system (benzene, pyridine, etc.) The numbering rules for benzene derivatives is the same as in cycloalkanes. The numbering systems for other aromatics depends on the accepted convention for each system.

5. Substituents are always ordered alphabetically (disregarding all prefixes except iso). Substituent names end in **o** (chloro, cyano), **y** (methoxy, carboxy) or **yl** (methyl, **phenyl** for benzene, cyclohexyl, 2-propenyl). Substituents on cyclic systems are **numbered** (except for a few heterocyclic and aromatic systems) from the point of **attachment** to the main molecular fragment. With these exceptions, substituents are named in the same manner as the parent molecule.

6. Double bonds are designated **E** or **Z** if each carbon of the double bond has **different** substituents. Double bonds with **two** substituents may be designated **cis** or **trans**. Cyclic systems with **two** substituents on different carbons are designated **cis** or **trans**. Chiral carbons are designated **R** or **S**. The cis-trans, (E, Z) or R, S) designations are placed at the **front** of the name of the molecule (or in front of the name of a substituent if they pertain to the substituent) and numbered if more than one.

7. Remember, when naming a molecule or drawing one from the name, always work **backwards**, starting at the end of the name containing the parent chain or cycle and working forward.

I.   Alkanes

A.   Find the longest continuous chain (do not include carbons of attached ring systems).

    1.   If naming an amine, ether, or sulfide as an alkane, pretend the N, O, or S is a carbon when counting the number of carbons for the chain, then name the N, O, or S as a substituent, as an **aza** (N), **oxa** (O), or **thia** (S).

B.   Name the chain as a methane, ethane, etc., placing this at the end of the name.

C.   Order the substituents alphabetically in front of the parent molecule name, regardless of their position number. If two or more of the substituents are the same indicate this by using the prefixes di-, tri-, tetra-, etc. in front of the substituent name. However, disregard the prefixes when determining the alphabetical order.

D.   Number the substituents so that they have the **lowest** possible **number combination**. That is, add up the numbers for all substituents from the two possible directions, and use the total number that is smaller. The numbers are placed in front of their respective substituent name-separated by hyphens. If more than one substituent is the same use position numbers, separated by commas to indicate the positions of each identical substituent.

E.   Use **hyphens** only to separate a **number** from **words**.

F.   Use **commas** only to separate **numbers** pertaining to the same group(s).

G.   Use **periods** to separate numbers in brackets for multi cyclic systems.

H.   Use **parentheses** to surround complex substituents if a number precedes a number within the name of the substituent.

I.   Use **brackets** if needed around substituents which already have parentheses. Brackets and parentheses are used to clarify. Brackets are also used in multi- cyclic systems (see below).

II   Cycloalkanes

A.   Determine the number of carbons in the ring; name as you would an alkane, but with the prefix **cyclo**.

B.   1.   Choose a position on the ring with a substituent to be Carbon #1 and proceed around the ring in the direction which results in substituents having the lowest possible number combination.

2. If a heteroatom (N, O, S) is replacing a carbon, it is **usually** assigned position #1.

3. Order the substituents alphabetically, disregarding prefixes, and write the position numbers in front of each substituent as was done for open chain alkanes.

C. If there are only two substituents and they are attached to different carbons the name is preceded by cis- if both are on the same side of the plane of the ring or trans- if on opposite sides.

D. For one or more substituents, the spacial orientation of the substituents can be better assigned by **R** or **S** if the carbons are chiral (see Section VI).

III. Aromatics

A. Benzenes are named as benzene preceded by the names of the substituents, ordered alphabetically, disregarding prefixes, and numbered in the same manner as with cycloalkanes.

B. If there are only two substituents, they can be said to be ortho (prefix o-) if located 1,2 on the ring, meta (prefix m-) if 1,3 and para (prefix p-) if 1,4. However, this convention has fallen into disfavor and should be avoided.

C. Two or more substituents can, and three or more must be numbered to indicate the position on the ring.

D. 1. **Other aromatics** are named and numbered as outlined in your text. The most common are napthalene, anthracene, phenanthrene, pyridine, quinoline, isoquinoline, pyrrole, furan, thiophene (see the examples below). Note that only the carbons which can be bonded to hydrogen are numbered.

2. Aromatics which have heteroatoms (N, O, S) can be named as aza, oxa, or thia aromatics if they have the same framework as a carbon system such as benzene (azabenzene = pyridine) napthalene (1-azanaphalene = quinoline), etc. (see the examples below). Heteroatoms are **usually** assigned position #1 (one exception is isoquinoline, 2-azanaphthalene).

IV. Alkenes and Alkynes

A. Find the longest chain containing the multiple bond(s) and give the lowest possible number to them, starting from one end of the chain.

B. The family name ending for alkenes is **ene**, the family name ending for alkynes is **yne**.

C.   The number of multiple bonds (above one) are designated by di, tri, etc.

D.   If both double and triple bonds are present, use the ending enyne and each is numbered directly in front of the suffix. Number from the end that gives the lowest numbered multiple bond. If numbering is the same, the alkyne takes precedence

E.   Orientation about a double bond is designated E or Z, with the sizes of the groups determined by the Cahn-Ingold-Prelog rules (Section VI). If the largest groups attached to each of the carbons of the double bond are on the same side of the line drawn through the double bond, the system is designated **Z** (Zusammen - German for together) if they are on the opposite side the system is designated **E** (Entgegen-German for on the opposite side). The E or Z is placed in front of the name or group if in a substituent-in parentheses and numbered if there are two or more E or Z double bonds in the molecule or substituent.

F.   Common nomenclature uses the prefixes *cis* and *trans*, instead of Z and E, respectively. The common nomenclature system is normally only be used if there are only two substituents on the double bond.

G.   Cyclic unsaturated systems always designate one of the multiple bond atoms as position #1, with the numbering proceeding **through** the double bond(s).

V.   Substituents

A.   **Simple**: Simple substituents include -Cl (chloro-), -Br (bromo-), -F (fluoro-), -I (iodo-), $-NO_2$ (nitro-), $-CH_3$ (methyl-), $-CH_2CH_3$ (ethyl-), $-C_6H_5$ (phenyl-), -OH (hydroxy-), $-OCH_3$ (methoxy-), $-OCH_2CH_3$ or -OEt (ethoxy-), $-OC_6H_5$ (phenoxy-), -COOH (carboxy-), $-OCOCH_3$ or -OAc (acetoxy-), $-C_6H_{11}$ (cyclohexyl-), etc. Also if there is an =O attached to the carbon (C=O), it is an **oxo**.

B.   **Complex aliphatic**: To name a carbon chain as a substituent, find the longest continuous chain attached to the parent chain or ring and name as such with -yl replacing -ane. The point of attachment of the substituent to the parent chain is **always** position #1 of the substituent.

1.   Give the lowest possible number to the point of attachment to the parent chain. This number precedes the name of the alkyl group.

2.   Substituents are given the lowest possible number and ordered alphabetically.

3.   Parentheses and/or brackets may be needed to insure the <u>entire</u> substituent is considered as one whole entity and its numbers are not confused with other numbers in the molecule.

4.   Substituents may in turn have substituents which are named, numbered and ordered as usual.

5.   The IUPAC system also allows common names for alkyl substituents. The following illustrate both systematic and approved common names: 1-methylethyl or isopropyl, 1-methylpropyl or *sec*-butyl, 2-methylpropyl or isobutyl, 1,1-dimethyethyl or *tert*-butyl, 2,2-dimethylpropyl or neopentyl.

C.   **Complex cycloalkyls**:

1.   To name a cycloalkane as a substituent count the number of carbons in the cycle and name by replacing **ane** with **yl**.

2.   Substituents and numbers are handled as above, remembering that the point of attachment of the cycloalkane to the parent system is always position #1, except for heterocyclic systems where the heteroatom is **usually** assigned position #1 (see the examples).

D.   **Complex aromatic**:

1.   Name the aromatic system (phenyl for all benzene derivatives, naphthyl for naphthalenes, pyridyl for pyridine, etc. See the examples).

2.   For phenyl the point of attachment is position #1. For other aromatics the point of attachment is determined by the numbering system of the aromatic ring (see examples).

E.   **Unsaturated systems**:

1.   To name an alkene or alkyne as a substituent, name as you would a alkene or cycloalkyne, replacing the ending **e** with yl and numbering from the **point of attachment**, indicating the position(s) of the double bond(s) with number(s).

VI.   **Optical Isomerism, Chiral Carbons**

A.   **Chiral Carbons** are designated R or S by the Cahn-Ingold-Prelog system.

B.   R or S is placed at the front of the name (or substituent if not part of the main chain or cycle) in **parentheses** and numbered if more than one, i.e. (4R, 5S, 7R).

C.   The sign of rotation is placed in parentheses at the beginning of the name (+)-, (-)-, (±)-.

D. **Cahn-Ingold-Prelog System.** A chiral carbon is designated as **R** or **S** by first determining the relative size of the 4 different groups attached to the carbon. Relative size is determined by comparing the atomic weight of each **element** directly attached to the carbon and assigning a priority to them of 1 to 4 with 1 being the largest-4 the smallest atom. If two or more atoms are identical, proceed in turn to the next atoms and determine the relative priorities. If atoms are connected by multiple bonds, consider **each bond** of the multiple bond to be separate bonds between elements. Continue from atom to atom until a difference in priorities is established. Next, group #4 (the smallest) is assumed to be the steering column of a steering wheel and the other three groups points on the steering wheel. If the wheel would have to be turned right to turn from group #1 to groups #2 and #3, the carbon is **R** (**R**ectus, Latin for right). If you must turn left, the carbon is **S** (**S**inister, Latin for left).

EXAMPLES

2-cyano-3-hydroxy-4-(1-methylethyl)heptane

3-chloro-4,4-dimethylcyclohexene

trans-1-bromo-3-phenylcyclopentane or
(1R,3R)-1-bromo-3-phenylcyclopentane

1-carboxy-4-ethyl-3-methoxybenzene
or 4-ethyl-3-methoxybenzoic acid

20

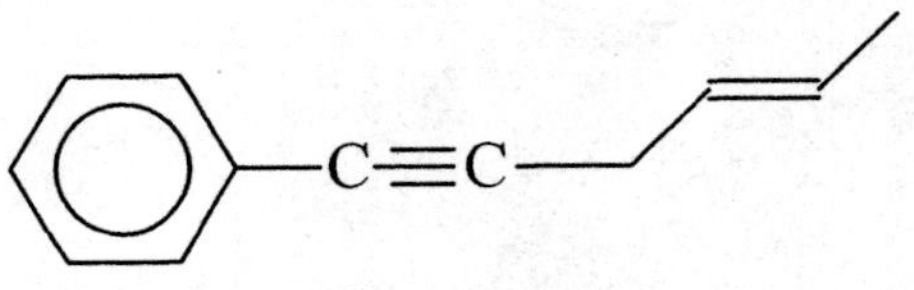

E-1-phenylhex-4-en-5-yne

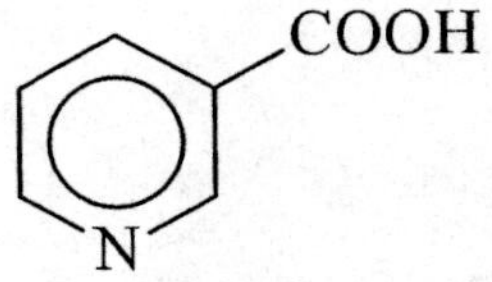

3-carboxypyridine or
3-carboxyazabenzene
or nicotinic acid

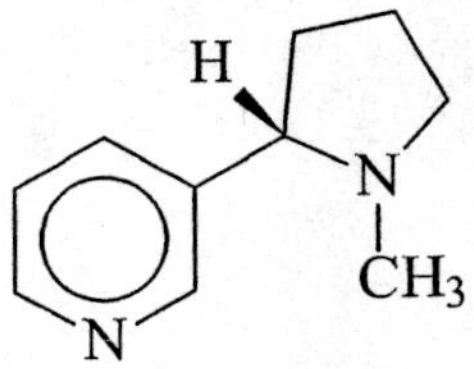

(S)-3-(1-methyl-2-pyrolidenyl) pyridine or
(S)-1-methyl-2-(3-pyridyl) pyrrolidine or
(S)-3-[2-(1-methylazacyclopentyl)] azabenzene
(Nicotine)

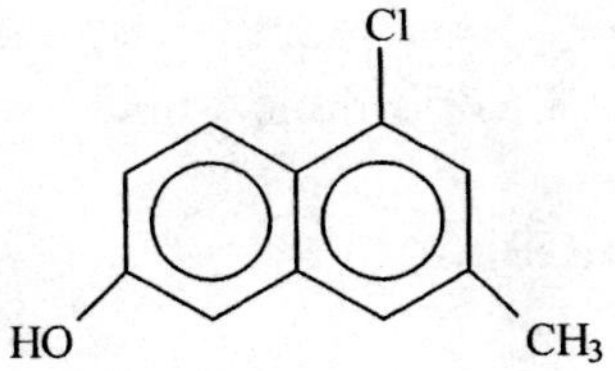

1-chloro-6-hydroxy-3-
methylnaphthalene

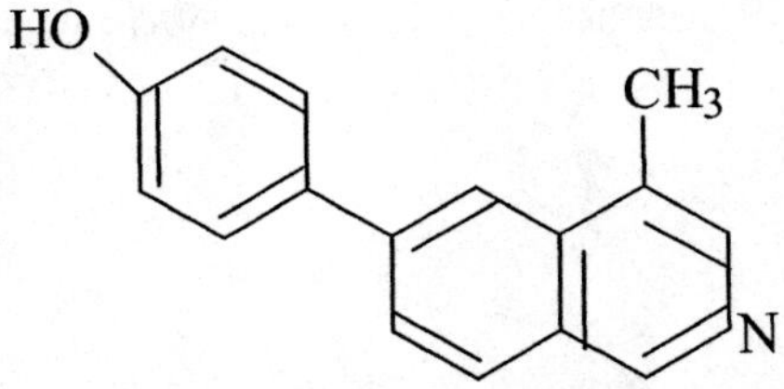

3-methyl-5-(4-hydroxyphenyl)isoquinoline or
2-aza-3-methyl-5-(4-hydroxyphenyl)naphthalene

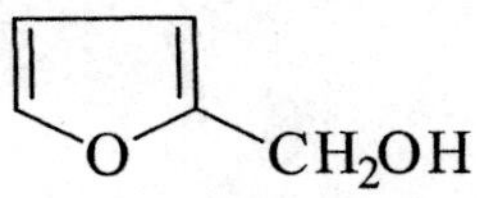

2-hydroxymethylfuran
2-hydroxymethyloxocyclopenta-2,4-diene

Organic Chemistry Reactions

A. Definitions

1.   Electronegativity

Electronegativity is a measure of the relative attraction that an atom has for the electrons it shares with another atom in a covalent bond.

2.   Non-polar Covalent Bond

A covalent bond is considered to be non-polar if the electronegativity difference between the two atoms is < 0.5 and their atomic radii are similar. (i.e. C-C, C-H, O-O)

3.   Polar Covalent Bond

A covalent bond is considered to be polar if the electronegativity differences between the two atoms is approximately > 0.5 < 1.9 or if the atomic radii of one of the atoms is much larger. (i.e. C-Cl, C-O, H-O, C-I)

4.   Ionic Bond

A bond is considered to be ionic if the electronegativity difference between the two atoms is approximately > 1.9. (i.e. $Na^+ Cl^-$, $Ca^{+2} SO_4^{2-}$)

5.   Carbocation

A carbocation is a carbon which has a positive charge (i.e. H_3C^+)

6.   Carbanion

A carbanion is a carbon which holds a negative charge (i.e. H_3C^-)

7.   Radical

A radical (sometimes called a free redical) is an atom which has an unpaired electron. (i.e. Br^-, H_3C^-)

8.   Charge on an Atom

An atom has a positive or negative charge when the number of electrons "owned" by the atom are less than or more than, respectively, the number of electrons the atom has as a neutral atom (the atomic number of the atom). An atom in a molecule is assumed to "own" all of its unshared electrons and one electron in each covalent bond. The charge is calculated by subtracting the number of electrons the atom "owns" in its outer shell from the number of electrons the atom has in its outer shell as a neutral atom. (i.e. H_3C, 4-3 = +1; $H_3C\bullet$, 4-4 = 0; H-O, 6-7 = -1)

9. **Bronsted-Lowry Acid**

A Bronsted-Lowry Acid (i.e. H-Cl) donates a H^+ to a Bronsted-Lowry Base.

10. **Bronsted-Lowry Base**

A Bronsted-Lowry Base (i.e. OH^-) accepts a H^+ from a Bronsted-Lowry Acid.

11. **Lewis Acid**

A Lewis acid is an atom that is electron deficient and can accept electron density from a Lewis base. The atom usually does not have a filled outer electron shell or is bonded to an electronegative atom.

Common Lewis Acids
a) H^+ (or H_3O^+)
b) R^+ (or any atom with a positive charge)
c) $R\bullet$ (or any atom with an unpaired electron, a radical)
d) AlX_3 and BX_3 (Al and B do not have an octet of electrons)
e) R-X (X is more electronegative or much larger than R)

12. **Lewis Base**

A Lewis base is an electron rich atom or bond which can donate electron density to a Lewis acid

Common Lewis Bases
a) O , N , S (neutral or negatively charged)
b) R^- (any negatively charged atom)
c) π Bond (double, triple bonds, aromatic rings)

13. **Electrophile**

Electrophile (Greek, electron-loving) is a term sometimes used rather than Lewis acid.

14. **Nucleophile**

   Nucleophile (Greek, Nucleus-loving) is a term sometimes used rather than Lewis base.

15. **Symbols**

   a) **R, Ar**   R represents 1 or more carbons, along with their associated atoms, attached to another atom. (i.e. H_3C-Cl can be represented by R-Cl)
   Ar represents an aromatic ring attached to another atom (i.e. Ar-OH)

   b) **X**   X usually represents a halogen atom (F, Cl, Br, I). Sometimes X is used to represent O, N or S plus attached atoms.

   c) ⟶   A straight arrow is used to indicate the conversion of reactants to products.

   d) ⇌   Two arrows are used to indicate a chemical reaction is reversible.

   e) ⟷   A two-headed straight arrow is used to indicate the structures connected are resonance structures.

   f)   A double-hook, curved arrow indicates the movement or sharing of 2 electrons from the electron rich Lewis base to the electron deficient Lewis acid, resulting in forming or breaking a covalent bond.

   g)   A single-hook curved arrow indicates the movement or sharing of 1 electron between two atoms or molecules.

   h)   The dipole arrow indicates the bond between two atoms is polar, with the arrow pointing toward the more electronegative atom or atom which holds more electron density in a bond.

   i) $\delta+, \delta-$ Another way to represent a polar bond between two atoms. The $\delta+$ means the atom has a partial positive charge or is electron deficient compared to the atom with the $\delta-$ symbol.

   j) **$h\nu, \Delta$**   These symbols represent energy. The $h\nu$ represents light energy (photons) and is derived from the formula $E = h\nu$, and h is Planck's constant and the Greek letter ν (nu) is the frequency of the light. The Greek letter Δ (delta) represents heat (the tip of a flame).

24

B. The Basic Reactions of Organic Chemistry

The most common reactions for organic chemicals fall into the following categories:
1.   Homolytic Radical Reactions
2.   Addition Reactions
3.   Elimination reactions
4.   Substitution Reactions
5.   Oxidation Reactions
6.   Reduction reactions
7.   Rearrangement Reactions

Some reactions (i.e. those of carbonyl compounds) involve mechanisms which are combinations of the basic reactions (i.e. addition and elimination).

1.   Homolytic Radical Reactions

A homolytic radical reaction differs from all other organic reactions because the reaction does not formally involve a Lewis base sharing electron density with a Lewis acid. A homolytic radical bond cleavage occurs when energy (usually heat or light) is absorbed by the bond. A homolytic radical bond formation occurs when two radicals collide.

Examples:

a) $Br_2 \xrightarrow{hv} 2\ Br\cdot$

b) $H_3C\cdot \ + \ \cdot CH_3 \longrightarrow CH_3 - CH_3$

2.   Addition Reactions

An addition reaction is the addition of a molecule A-B (i.e. H-Cl or H-OH) to a double or triple bond.

Examples:

a)

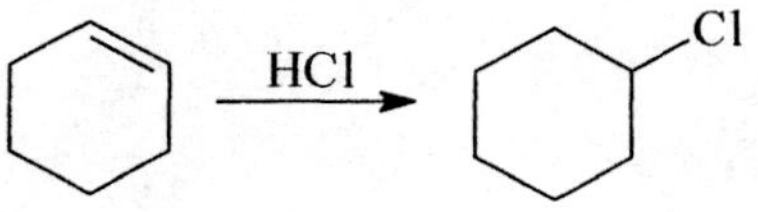

b)

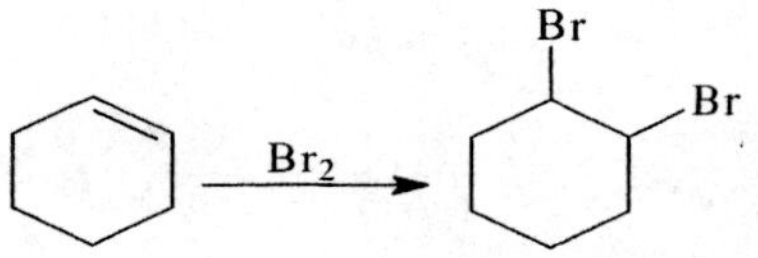

c) 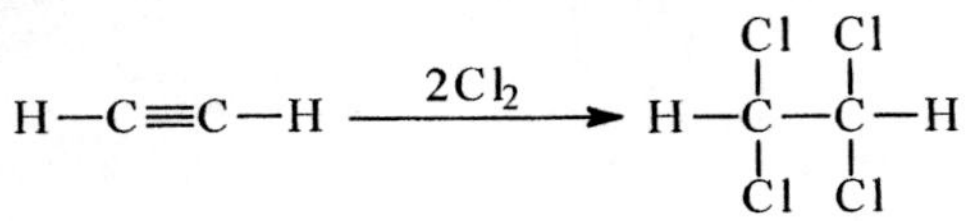 $H-C\equiv C-H$ $\xrightarrow{2Cl_2}$ Cl-C-C-H (structure showing $H-\underset{\underset{Cl}{|}}{\overset{\overset{Cl}{|}}{C}}-\underset{\underset{Cl}{|}}{\overset{\overset{Cl}{|}}{C}}-H$)

3.   Elimination Reactions

An elimination reaction is the removal of a molecule A-B (i.e. H-Br, H-OH) from two adjacent atoms to form a double or triple bond.

Examples:

a)

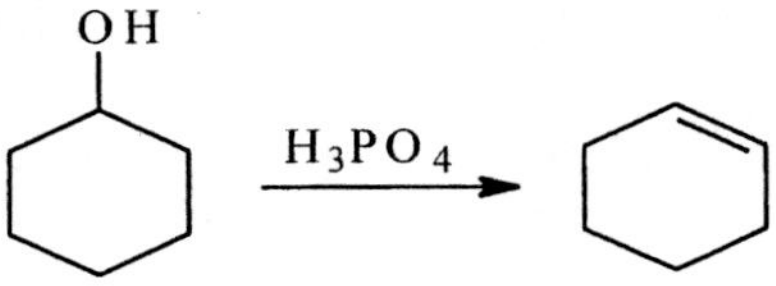

$\xrightarrow{H_3PO_4}$

b)

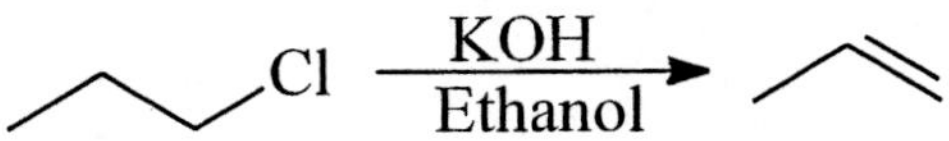

$\xrightarrow[\text{Ethanol}]{\text{KOH}}$

c)

$\xrightarrow{2NaNH_2}$

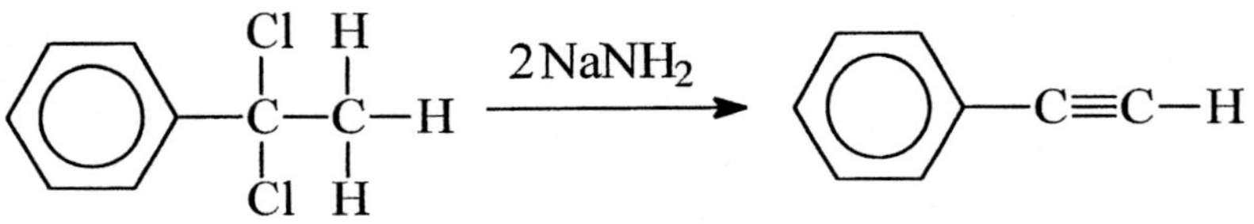

4.   Substitution Reactions

A substitution reaction is the replacement of a group (X) by another group (Y). Three common substitution reactions occur.

a)   Free Radical Halogenation of Alkanes
b)   Nucleophilic Substitution
c)   Electrophilic Aromatic Substitution

Examples:

a) 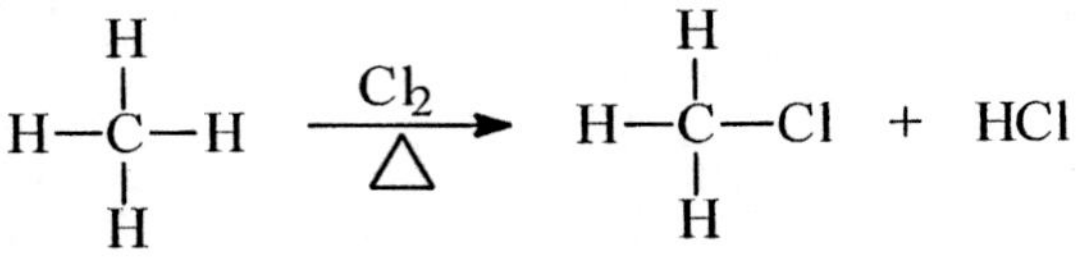 $H-\underset{\underset{H}{|}}{\overset{\overset{H}{|}}{C}}-H$ $\xrightarrow[\triangle]{Cl_2}$ $H-\underset{\underset{H}{|}}{\overset{\overset{H}{|}}{C}}-Cl$ + HCl

b)

$\xrightarrow{CN^-}$

c)

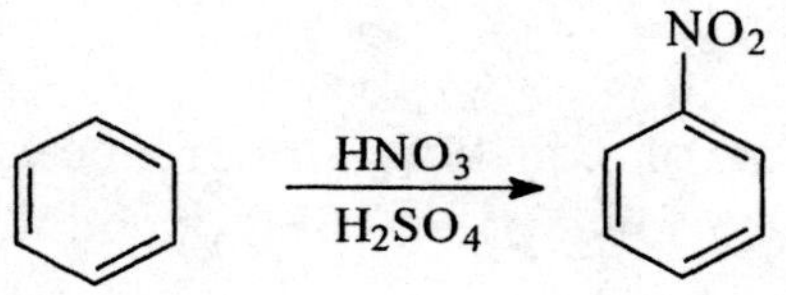

5. Oxidation Reactions

In most cases, oxidation in organic reactions involves the addition of an oxygen atom, removal of a molecule of hydrogen (H-H) or removal of an electron from an atom or molecule.

Examples:

a) 1° Alcohols

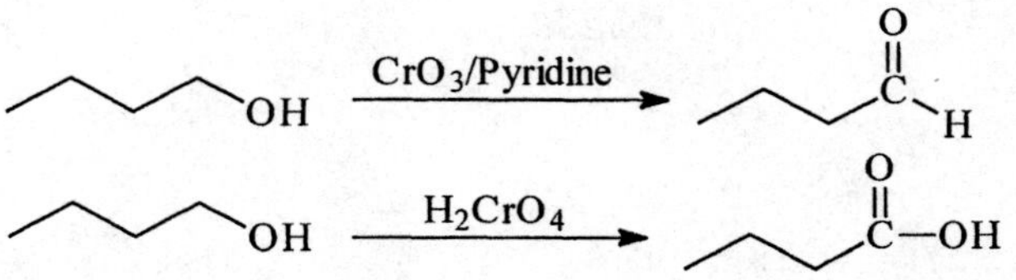

b) 2° Alcohols

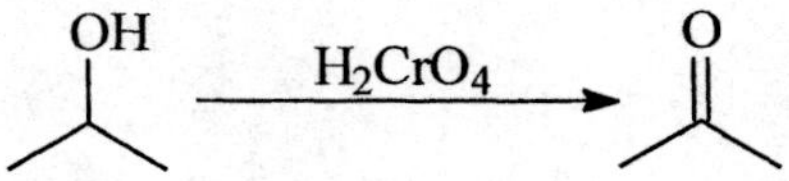

c) 3° Alcohols

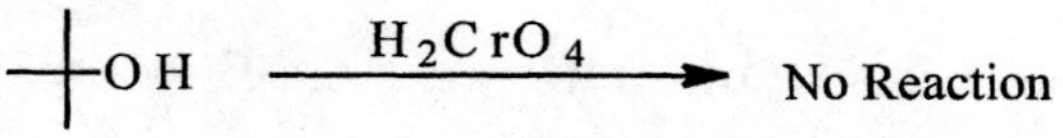

d) Double Bonds

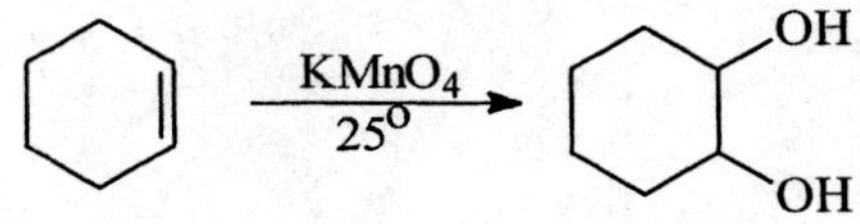

e) Aromatic Side Chains

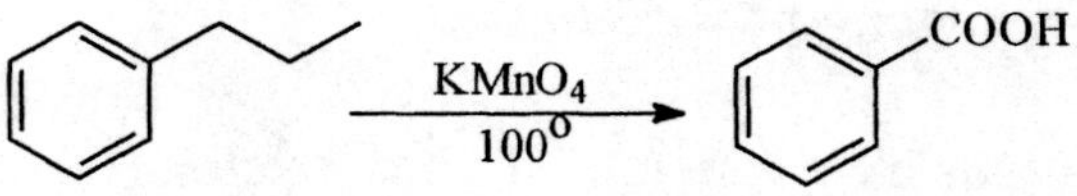

f) Complete Combustion

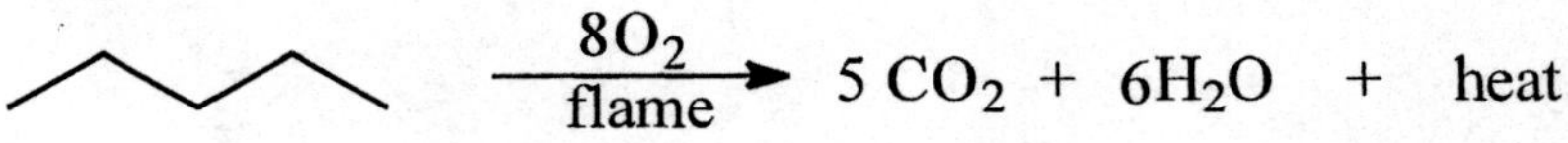

$$\xrightarrow[\text{flame}]{8O_2} \quad 5\,CO_2 + 6H_2O + \text{heat}$$

6. Reduction Reactions

Reduction reactions are just the opposite of oxidation. Thus, reduction is the removal of an atom of oxygen, addition of a molecule of hydrogen (H-H) or addition of an electron to an atom or molecule.

Examples:

a) Carbon Halogen Bonds

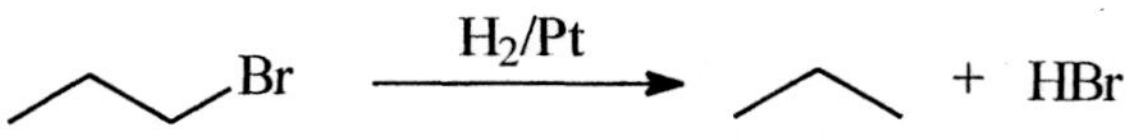

b) Double Bonds

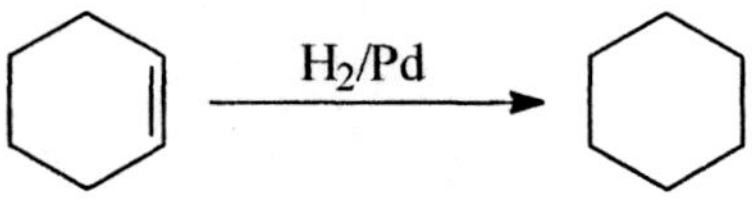

c) Triple Bonds

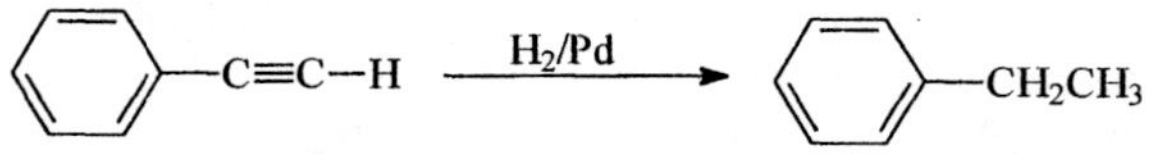

d) Aldehydes

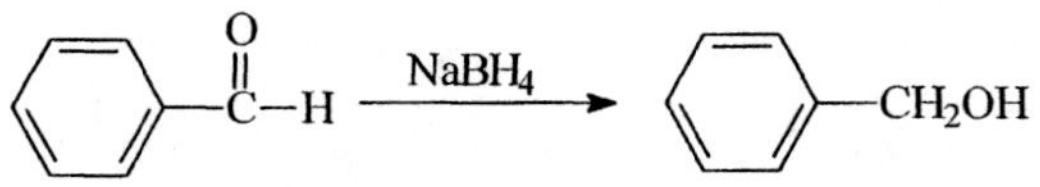

e) Ketones

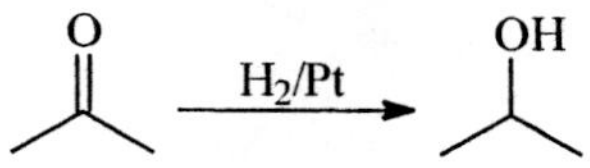

f) Carboxylic Acids and Acid Derivatives

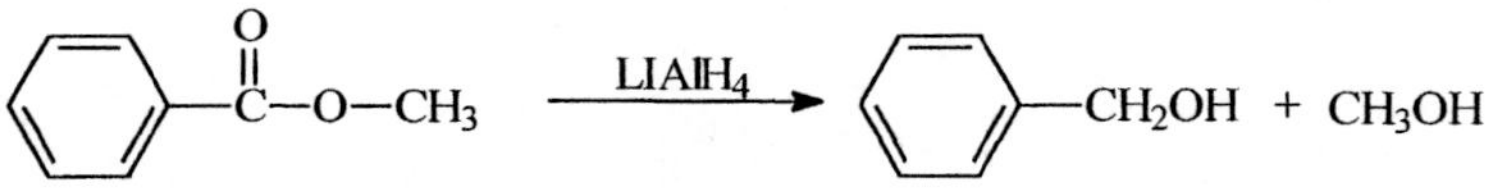

g) Nitro Compounds

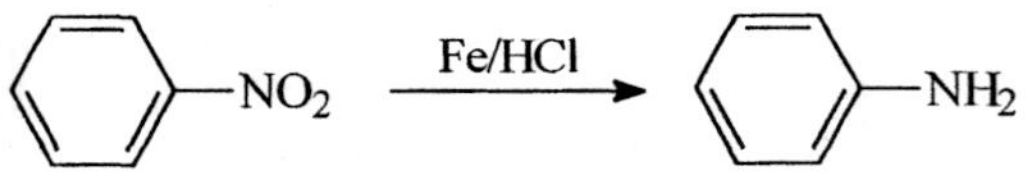

7. Rearrangement Reactions

Carbocations may undergo carbon skeleton rearrangements to give a more stable carbocation. The relative stabilities of carbocations are benzyl, vinyl >> 3° > 2° > 1°.

Example:

a)

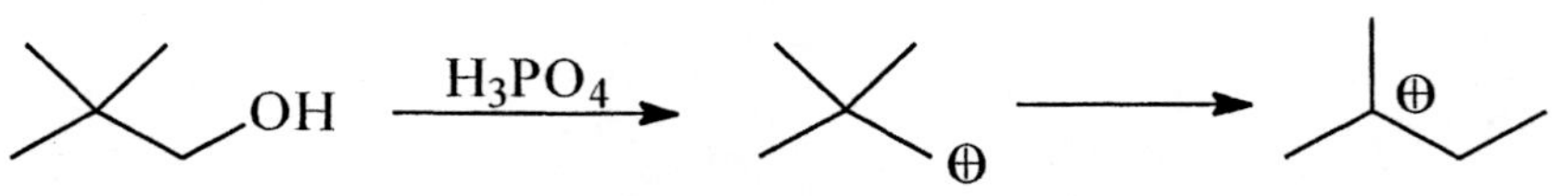

C. Reaction Mechanisms

Most organic chemistry reactions involve a Lewis base sharing two electrons with a Lewis acid, as in the following example of water reacting with hydrogen chloride gas:

$$H_2O: \curvearrowright H \overset{\frown}{\longrightarrow} Cl \longrightarrow H_3O^{\oplus} + Cl^{\ominus}$$

The two-headed curved arrow starting at oxygen is used to represent the Lewis base (the oxygen) sharing 2 electrons with the Lewis acid (the hydrogen attached to the more electronegative chlorine). Since electrons are shared by the two atoms, a bond is formed between them. Also, since the neutral oxygen is giving electron density to the positive hydrogen, oxygen in the product has a positive charge and hydrogen is neutral. The H-Cl bond is broken during this reaction, since hydrogen can normally have only one bond (period I in the periodic table). A good rule of thumb that is usually true is **"make a bond, break a bond."**. When the H-Cl bond breaks, the 2 electrons in the bond move toward chlorine, since Cl is the more electronegative atom. Note that the total charge on the starting materials is 0 and the total charge on the products is 0. Charge does not change during a reaction.

Remember, Lewis bases are electron rich (can share electron density) and Lewis acids are electron deficient (can accept electron density). Most Lewis acids (such as H^+, Br^+ or the boron in BH_3) need two additional electrons to have a completed outer electron shell (2 or 8 electrons). Therefore, reactions involving these Lewis acids involve the sharing of 2 electrons and thus a two-headed curved arrow.

Some reactions (radical ractions) involve atoms which have an unpaired electron and thus need 1 additional electron to have a complete electron shell. These radicals accept a single electron from an atom or chemical bond. Since radical reactions involve the movement of 1 electron, a single-headed curved arrow is used in these reactions. In the following reaction, the bromine atom has an unpaired electron and needs 1 more electron to have a filled outer electron shell (an octet), and thus is the Lewis acid. One of the 2 electrons in the C-H bond is shared with the Br. The second electron in the C-H bond remains on the carbon. Thus a total of 3 electrons are involved in this reaction and thus 3 single-headed curved arrows are used to show this chemical mechanism. Note that there is 1 unpaired electron for the starting materials and 1 unpaired electron for the products. The number of unpaired electrons does not change in a reaction.

In a few cases, radical reactions do not involve a formal Lewis acid or Lewis base. These so-called homolytic (one molecule) reactions occur when enough energy (in the form of heat or light) is available to break the bond between two atoms. One electron stays with each atom and 2 radicals are thus formed. These radicals are now electron deficient Lewis acids which undergo normal Lewis acid-base reactions.

One last case involves two radicals sharing their unpaired electrons to reform a bond. Again, there is no formal Lewis acid or base in this reaction.

D. Reaction Mechanism Examples

1. Free Radical Halogenation

2. Addition Reactions

a.

b.

3. Elimination Reactions

a.

b.

4. Substitution Reactions

a.

b.

c.

5. Hydrolysis of an Ester (Addition + Elimination)

Experiment 1
An Introduction to Organic Chemistry

Carbon is unique among the elements in that it combines with other carbons, as well as other elements, to form an almost infinite variety of chain and cyclic covalently bonded molecules. Carbon has four valence electrons which, in stable molecules, are allowed to form covalent bonds with sp, sp^2, or sp^3 hybridization of the carbon atomic orbitals. Thus carbon can have a coordination number of 2 (with a single and triple bond and a linear geometry), 3 (with two single and one double bond and a trigonal planar geometry) or 4 (with four single bonds and a tetrahedral geometry).

Carbon molecules are often represented in your text as flat structures with sticks between the atoms representing the molecular orbitals (bonds) holding the molecule together. Below are some examples (the names in parentheses are non-systematic names):

Ethene (Ethylene)	Methane	Trichloromethane (Chloroform)	Methanal (Formaldehyde)

Benzene	Ethyne (Acetylene)	Methylamine	Hydrogen Cyanide

To save space these structures are often written with all atoms in a line and the number of atoms of each type indicated by number with subscripts, i.e., CH_4, $CHCl_3$, H_2O, CH_3CO_2H, C_6H_6, C_2H_2, CH_3NH_2, HCN.

In fact these structures do not represent accurately the three dimensional geometry of the molecules, an obvious difficulty since the paper is in two dimensions. We therefore use conventional representations to indicate three dimensions on paper. In a three-dimensional representation when a bond is written as a solid stick (i.e., H-C) this means it is in the plane of

33

the paper. If the bond is coming out toward you write the bond as a wedge, the larger end toward you, and if the bond is going away from you write it as a dotted line. Thus we have bromochloroiodomethane:

$$\text{Cl} \diagdown \underset{\underset{\displaystyle I}{\diagup}}{\overset{\displaystyle Br}{\underset{\displaystyle H}{C}}}$$

Note that the carbon, when possible, is placed in the middle and for a tetrahedral geometry two other elements attached to carbon are also in the plane of the paper with the carbon. Thus, the previous molecules can be written (the hybridization of the C, N, O atoms has also been indicated) in three dimensions as:

$$\underset{\displaystyle sp^3}{\overset{\displaystyle H \diagdown \; \diagup H}{H \diagup C \diagdown H}} \qquad \underset{\displaystyle sp^3}{\overset{\displaystyle H \diagdown \; \diagup Cl}{Cl \diagup C \diagdown Cl}} \qquad \underset{\displaystyle sp^3 \;\; sp^3}{\overset{\displaystyle H \diagdown \; \diagup H}{H \diagup C \diagdown NH_2}} \qquad \underset{\displaystyle sp^2}{\overset{\displaystyle sp^2 \;\; O}{\overset{\|}{H \diagup C \diagdown H}}} \qquad \underset{\displaystyle sp^2}{\overset{\displaystyle H \diagdown \quad \diagup H}{H \diagup C=C \diagdown H}}$$

$$H-C\equiv C-H \qquad H-C\equiv N$$
$$\qquad sp \qquad\qquad\quad sp$$

Note that atoms attached directly to an sp^2 or sp hybridized atom are drawn in the plane of the paper. Note that nitrogen, oxygen, and sulfur may also be sp, sp^2, or sp^3 hybridized and that for C, N, O, S if the atom has a **triple bond** it is normally sp with a linear geometry, if it has a **double bond** it is normally sp^2 with a trigonal planar geometry and if it has only **single bonds** it is sp^3 with a tetrahedral geometry.

Note also that drawing a molecule **oriented differently** in space or by rotating atoms about single bonds **does not** make it a **different** molecule

$$\text{Thus} \quad \overset{\displaystyle Cl}{\underset{\displaystyle H}{H-C-H}} \quad \text{and} \quad \overset{\displaystyle H}{\underset{\displaystyle H}{H-C-Cl}} \quad \text{are the same molecules since they can be } \textbf{superimposed} \text{ (can}$$

be identically placed on top of one another) by rotating in space.

34

Thus H—C—C—C—H and [structure] are the same molecule because they

can be superimposed after rotation about the C—C bond. Normally the molecule is written the first way since it can be placed on one line. Cyclic alkanes are often written in three dimensions with the rings flat, i.e.,

even though, as we shall see when we build models, they are not (except for cyclopropane) actually flat and in the perspective view the groups attached to the carbons are not actually oriented straight up and down.

Several other features of molecular geometry will become apparent when we build the models, including E,Z (cis,trans) isomerism in alkenes, cis, trans isomerism in cycloalkanes and optical isomerism.

OBJECTIVES

The purpose of this experiment is to familiarize you with the three dimensional shapes of molecules and to enable you to relate the shapes to pictorial representations and vice versa. Build the models as indicated and answer the questions during the lab on a separate sheet of paper to hand in at the end of the lab period (keep this handout). Answer completely and with an explanation or drawing, not just a yes or no. If you do not understand the instructions, questions or conventions for drawing the molecules, do not ask another student for help, ask your instructor.

If you are using the **Prentice Hall plastic model kit**, the short sticks represent **all** single bonds, the long flexible sticks are used for **multiple** bonds, the **black** balls are carbons, the **white** balls are hydrogen, the **red** balls are oxygen, the **blue** balls are nitrogen, the **yellow** balls are sulfur and the **green** balls represent halogen (F, Cl, Br, I). If you need to represent individual halogens, use **green** for chlorine, **red** for bromine, **blue** for iodine in the following experiment. You will need to **share** kits with your partner to construct the larger molecules.

1. Build a model of methane

CH$_4$,   H—C—H (with H above and H below central C)

(tetrahedral representation with H, C, H arrangement)

Note that the hydrogens are at the corners of a tetrahedron with the carbon in the center.

2. Make CH$_3$Cl (Cl = **green** ball), CH$_2$ClBr (Br = **red** ball), CHClBrI (I = **blue** ball).

3. Build another model of CHClBrI identical to the first, now interchange any two atoms attached to the carbon of one model. Note that the two molecules cannot now be superimposed. If you look carefully, you'll see that the two molecules are mirror images of each other. These molecules are called isomers - to be exact, optical isomers. (**Molecules are called isomers if they have the same molecular formula, but have their atoms arranged in space differently and cannot be superimposed by rotation in space or about single bonds.** Optical isomers are non-superimposed by rotation in space or about single bonds, and also are non-superimposable **mirror images**). A molecule will have a non-superimposable mirror image (an optical isomer) if there is at least one carbon atom with 4 **different** groups attached to it (a so-called chiral, pronounced kyral, carbon), and if a plane of symmetry **cannot** be drawn through the molecule dividing it into two mirror halves.*

4. Build ethane, C$_2$H$_6$, CH$_3$CH$_3$

(structural formula of ethane with H—C—C—H and tetrahedral representation)

Note the free rotation about the C-C bond.

Does ethane have any chiral carbons?

Ethane has at least two planes of symmetry. Draw one plane which does <u>not</u> cut through the carbons and one plane which does cut through the atoms.

5. Build propane, C$_3$H$_8$, CH$_3$CH$_2$CH$_3$,

(structural formula of propane with H—C—C—C—H and tetrahedral representation)

Is there any other way to arrange the atoms, using all of them, that results in an isomer of propane?

*The 2 optical isomers (also called enantromers) are named with a prefix R or S, which indicates the absolute three-dimensional structure for each. We will not use R, S nomenclature in this experiment.

6. Build butane, C_4H_{10}, $CH_3CH_2CH_3$,

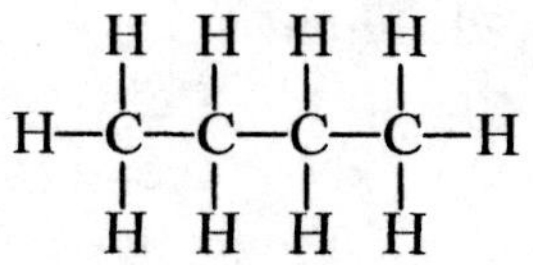

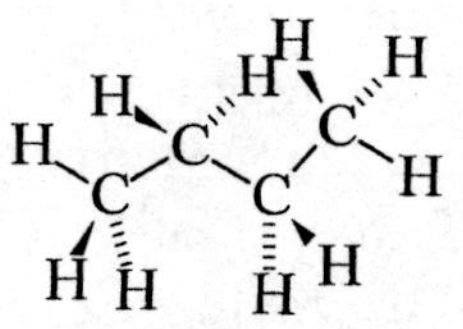

$$H-\overset{\displaystyle H}{\underset{\displaystyle H}{\overset{|}{\underset{|}{C}}}}-\overset{\displaystyle H}{\underset{\displaystyle H}{\overset{|}{\underset{|}{C}}}}-\overset{\displaystyle H}{\underset{\displaystyle H}{\overset{|}{\underset{|}{C}}}}-\overset{\displaystyle H}{\underset{\displaystyle H}{\overset{|}{\underset{|}{C}}}}-H$$

Rotate about the center C-C bond 180°. These two forms of the molecule may seem to be different, but are not. They are called rotational conformations or **rotomers**. They are not considered to be different molecules because rotation occurs so rapidly that they cannot be physically isolated before interconverting. How many possible rotomers for butane do you think there are?

7. Build 2-methylpropane (isobutane) by removing a hydrogen on the second carbon and exchanging it with the fourth carbon and three H's (CH_3, a methyl group)

C_4H_{10}, $CH_3CH(CH_3)_2$,

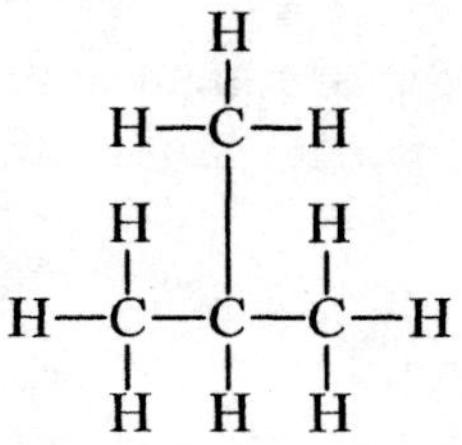

$$\begin{array}{c} H \\ | \\ H-C-H \\ H\quad | \quad H \\ H-C-C-C-H \\ | \quad | \quad | \\ H \quad H \quad H \end{array}$$

Isobutane is not a systematic or preferred name. It was named this way because it is an isomer (a structural or constitutional isomer) of butane since it contains the same number of atoms as butane but with the atoms arranged differently.

Are there any other isomers of butane?

8. Build pentane C_5H_{12} and all the possible isomers of pentane.

9. Build hexane, C_6H_{14}. Draw all the possible isomers of hexane (use flat structures).

10. Remove a H and stick from each of the terminal carbons of the hexane chain. Then connect the two carbons with a short stick. This is cyclohexane C_6H_{12}. Is the model flat? Is it flexible? There are two most stable conformations for cyclohexane: the so-called chair and boat forms, as shown below. Flex your model so it looks like each of the pictures.

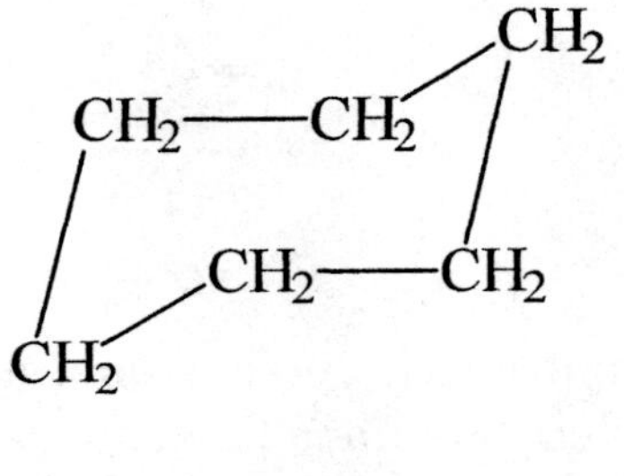

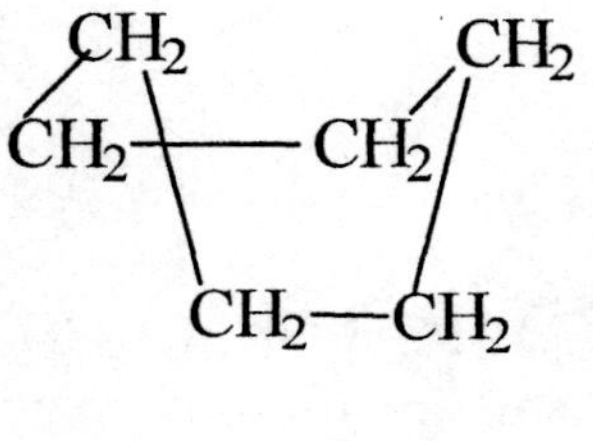

11. Conformations differ only in that they have different orientations about single bonds and at room temperature cannot be said to be different since they interconvert so fast that they cannot be isolated. The term conformation is always used for cyclic molecules and the term rotomers (which conformers really are) often used for open chain molecules. Note that when you lay the chair flat on the table, one hydrogen on the carbons raised higher is pointed straight up and the other slightly down (they are referred to as axial and equatorial hydrogens respectively). How are the hydrogens on the lower carbons oriented? Now twist the bonds so that the raised right end carbon (look at the picture) in the chair is down and the lower left hand carbon is up. Note that the equatorial and axial hydrogens become reversed when the chair forms are interconverted. This can be more easily seen if you replace any one of the axial - straight up or down - hydrogens with a green ball)? Ask your instructor for help if you need it!

Finally, note that the formula for cyclohexane (C_6H_{12}) has 2 fewer H's than that for hexane (C_6H_{14}) and that it is **not** an isomer of hexane.

12. Remove the 2 H's from any carbon and replace them with Cl's (green balls). If you do the same to any other carbon instead, have you formed an isomer or is it the same molecule. Hint, pretend the molecule is flat.

13. Again, pretending the molecule is flat, remove 2 H's from any 2 carbons from the **same** side of the plane of the molecule and replace by Cl. The molecule is said to be substituted **cis**. If you place the Cl's on opposite sides of the plane of the molecule, the molecule is said to be substituted **trans**. How many dichlorocyclohexane isomers are there? Draw them.

We can only use the cis-trans nomenclature for cyclic molecules which contain 2 substituents. Attachment of 3 or more substituents requires the use of R, S nomenclature.

14. Construct ethene (ethylene), C_2H_4, CH_2CH_2,
$$\begin{array}{cc} H & H \\ \diagdown & \diagup \\ & C=C \\ \diagup & \diagdown \\ H & H \end{array}$$

The double bond is constructed with the model kits by connecting the 2 carbons with two **flexible** plastic sticks. Note the atoms are all in the same plane and C substituted trigonally. Can you rotate about the C=C bond? The model should resemble the following:

15. Construct 2-butene, C_4H_8, $CH_3CH=CHCH_3$.
   Note that there are 2 isomers of 2-butene, with the terminal CH_3's on the same side (Z-2-butene, common name cis-2-butene) or opposite side (E-2-butene, common name trans-2-butene) of an

imaginary line drawn through the double bond, connecting the carbons. Construct, draw and name each. These are called **geometrical isomers**. The **Z** stands for the German word **Z**usammen - meaning "together" and the **E** stands for **E**ntgegen - meaning on the "opposite" side.

An older designation used the terms **cis** and **trans** respectively but this terminology has fallen out of favor. Cis-trans prefixes are normally used when there are only 2 substituents on the double bond. If there are 3 or 4 substituents (other than H), E, Z nomenclature is usually used.

Atoms which are bonded with a double bond are sp^2 hybridized and atoms are attached to them in trigonal planar geometry.

Please Note: The shapes of the molecular orbitals for double (and triple) bonds is **not accurately represented** by this type of model kit, although the bond angles are accurately represented. Your lecture text shows the proper shapes of the molecular orbitals.

16. Construct ethyne (acetylene), C_2H_2, H-C $\equiv$ C-H. The triple bond is constructed with these models by connecting the 2 carbons with 3 flexible plastic tubes. Note that the 4 atoms are arranged in a linear manner. The carbons are sp hybridized with a linear geometry, as is nitrogen when it is sp hybridized (H-C $\equiv$ N).

17. Construct benzene, C_6H_6,

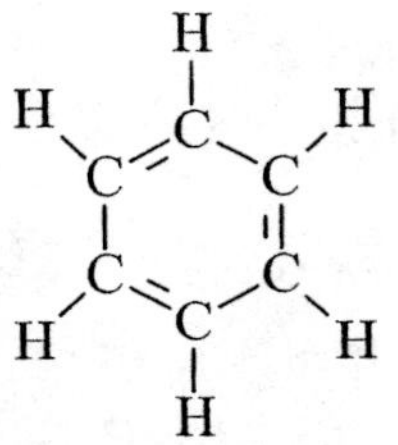

      The same method is used to construct other **aromatic** molecules like benzene such as naphthalene, toluene, pyridine, and anthracene. Look up the structures of these molecules in your lecture text, draw and name them.

What hybridization type do the carbon atoms have in benzene? The molecule appears to have three double bonds and three C-C single bonds but this is an artifact of a model which cannot illustrate the true electron density in the molecule. Rotate your model clockwise 60° so it looks like the following picture

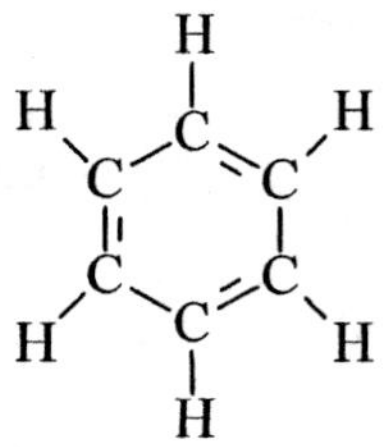

This and the original representation for benzene are not different molecules. The two written structures are called **Kekule** formulas and neither separately represents the true electron density distribution in benzene. Benzene is actually better represented as a **combination** of both structures - with neither C-C single or C=C double bonds but somewhere in between. Sometimes this is made clearer by drawing benzene as shown:

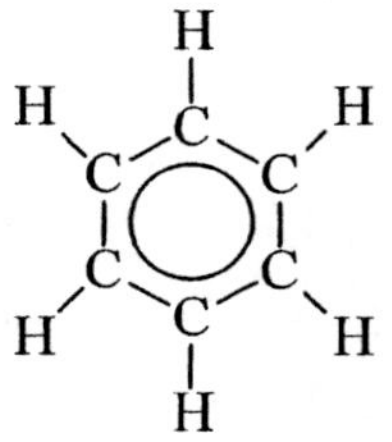

indicating that the electron density is equally distributed around the ring. This property is shared by all of the aromatic family of chemicals. The fact that **benzene has no double bonds** will be evident to you when you perform reactions with benzene derivatives and find that **they do not chemically react like molecules with double bonds**.

18. Remove any two H's from the benzene and replace with Cl's (green balls). How many dichlorobenzene **positional isomers** can you construct? Draw them also.

There is another method for drawing structures with **three** or more carbons often used to save time and for sake of clarity: The so-called skeleton structures. Essentially carbons are **not** written, only the **bonds** connecting them. H's attached to carbon are **not** written. However, all other atoms (N,O,Cl, Br,I,S) and the H's attached to them are drawn. H's attached to carbon are shown, however, if you want to emphasize three dimensions at the carbon. Also, the carbon is often written when attached by a multiple bond to O, N, or S and the H is **always**

shown on the carbon of an aldehyde, i.e., 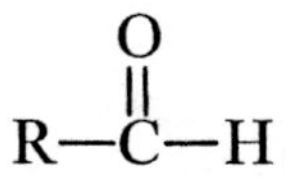 or

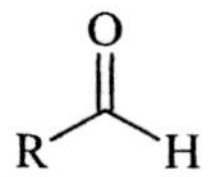

If the **C is written**, however, **H's attached must be written** as well.

40

Some of the conventions used above may be difficult for you to understand at this time. Ask your instructor for help if you have any trouble interpreting the structures.

Cyclic structures are also often written as though they were flat, as a regular polygon (i.e. a hexagon) **in perspective**, with the attached substituents drawn straight up and down. Thus:

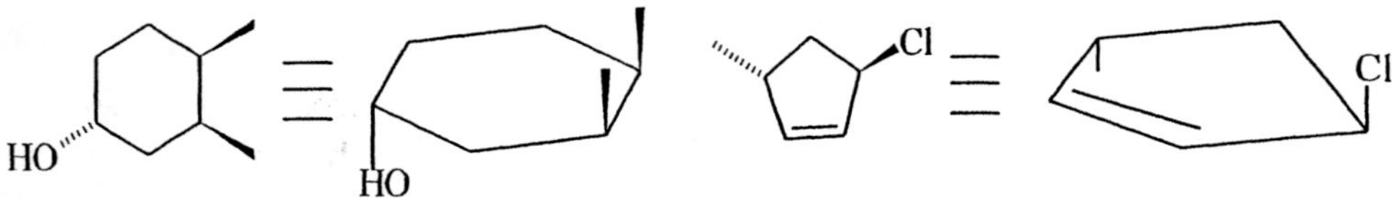

There are some other useful abbreviations and conventions you should become familiar with $-CH_3$ or -Me (from **M**ethyl) $-CH_2CH_3$ or-Et (from **E**thyl), $-CH_2CH_2CH_3$ or -Pr (from **P**ropyl)

$-CH_2CH_2CH_2CH_3$ or $-$Bu (from **B**utyl), $-C\equiv N$ or $-CN$, $-\overset{\overset{\displaystyle O}{\|}}{C}-OH$ or $-COOH$ or

$-CO_2H$, $-\overset{\overset{\displaystyle O}{\|}}{C}-H$ or $-CHO$, or 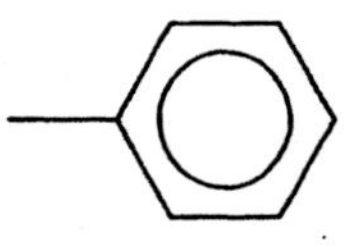 or $-$Ph (from phenyl)

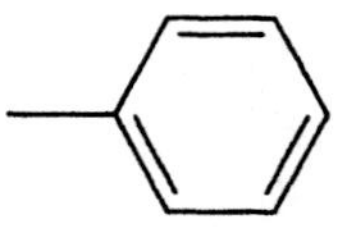

or ~ϕ (from the Greek Letter for f).

Note that the bonds connecting two atoms are sometimes omitted **if** it is **clear** how the atoms are bonded to one another.

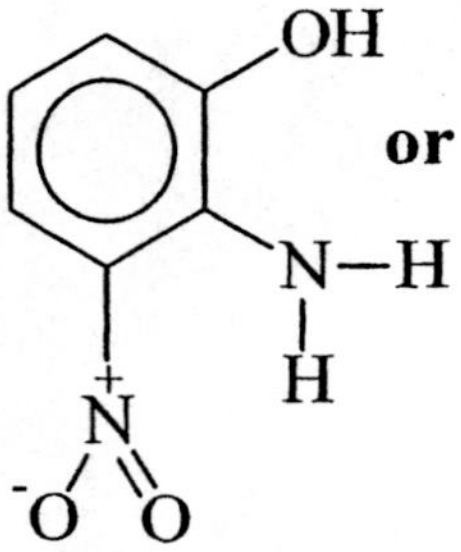

or 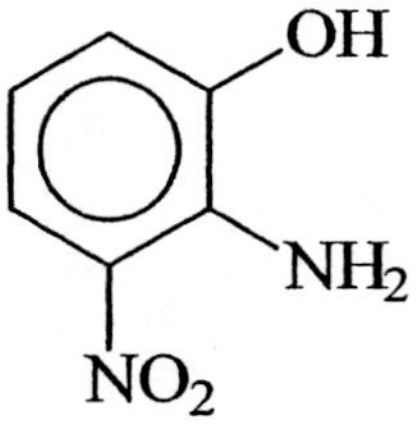 **but not** 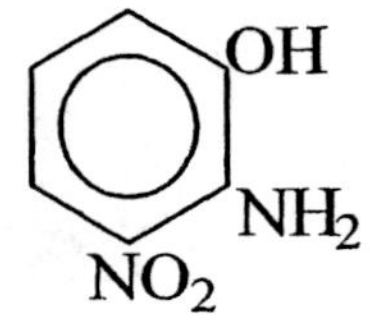 since the O and N's

may be interpreted by someone as being part of the ring system.

Remember also that **carbon normally has 4 bonds, nitrogen 3 bonds, oxygen and sulfur 2 bonds, the halogens and hydrogen 1 bond.**

Examples

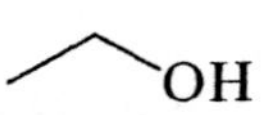

| ethanol | cyclohexene | propanone | 3-hexyne | 2-hydroxy-2-methylpropane |
| (ethyl alcohol) | | (acetone) | | (t-butyl alcohol) |

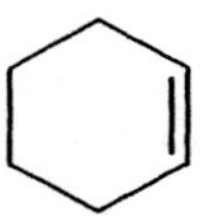

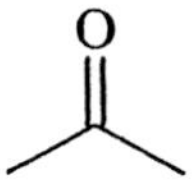

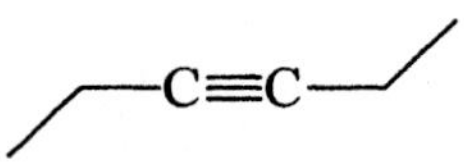

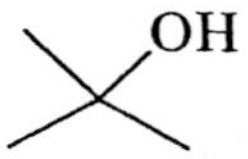

42

19.

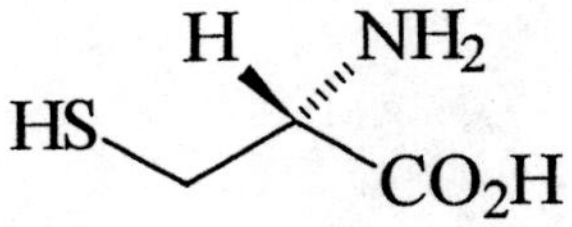

cysteine

This molecule is optically active. Put a little **star** at the **chiral** carbon. What are the 4 groups attached to the chiral carbon? Draw the optical isomer of cysteine. (cysteine is an essential amino acid).

20.

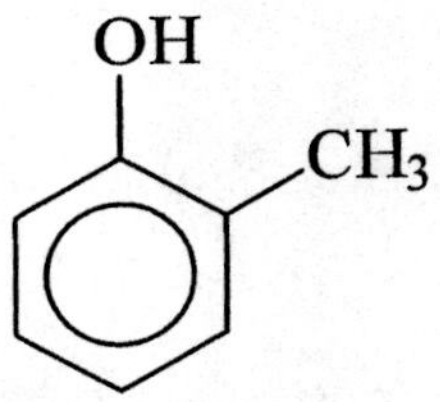

2-methylphenol
(o-cresol)

Draw as many of the isomers of o-cresol as you can. Remember, the isomer may have any structure, as long as it has the same molecular formula.

Experiment 2
Properties and Reactions of Hydrocarbons and Halogenated Hydrocarbons

Hydrocarbons are those gaseous, liquid and solid organic compounds which contain only carbon and hydrogen. Halogenated hydrocarbons have one or more of the hydrocarbon hydrogens replaced by fluorine, chlorine, bromine or iodine. The hydrocarbons consist of four families of compounds: the alkanes, the alkenes, the alkynes and the aromatic compounds. All other organic compounds, including the halogenated hydrocarbons, are derivatives of the hydrocarbons with one or more hydrogens or carbons replaced by atoms such as oxygen, nitrogen, sulfur, phosphorous, halogens or even metals such as lithium. In this experiment we will examine some of the more important chemical reactions and physical properties of hydrocarbons and chlorinated hydrocarbons. By chemical reactions, we mean an actual change in structure of the molecule, as when a molecule of bromine adds to an alkene. By properties, we mean things such as the tendency of the material to evaporate or to dissolve in a solvent. Dissolving is not a chemical reaction, since the material can be recovered unchanged. This should be distinguished from what occurs when a chemical reacts with a solvent to form another chemical species which then becomes soluble in the solvent. *Often, if there is a reaction, some visible change, such as a color change, occurs as well.* Please follow the instructions carefully, making changes only if your instructor tells you to. Since you will be working with a partner and there are many parts to this experiment, you and your partner will have to work as a team, dividing up the tasks. However, since both of you will be responsible for the observations that your partner makes, and vice versa, *it is necessary that you write down your observations as they occur during the lab* on the report form at the end of the experiment, so that you can go over the material after class and finish writing up the report. It is also recommended that you share your observations with your partner, as they occur, and that both partners participate in observations taken away from your desk, such as the flammability test.

I. Alkanes

Free Radical Halogenation

The hydrogens of alkanes are replaced or substituted with halogen (bromine or chlorine) atoms in the presence of heat or light, which initiates a free radical reaction. We will use bromine (a liquid at room temperature) dissolved in methylene chloride, a brown-orange solution, for the test.

II. Alkenes

Reaction with Br_2/CH_2Cl_2

Alkenes undergo **addition** reactions, in which a molecule, such as Br_2, adds to the double bond of the alkene and becomes part of the molecule, as indicated below. Recall that a double bond is actually composed of a **sigma bond and a pi bond**, which in turn is composed of two parallel overlapping **p atomic orbitals**. It is actually the two electrons in the pi bond which are reacting with the Br_2.

$$H_2C=CH_2 \xrightarrow{Br_2} Br-CH_2-CH_2-Br$$

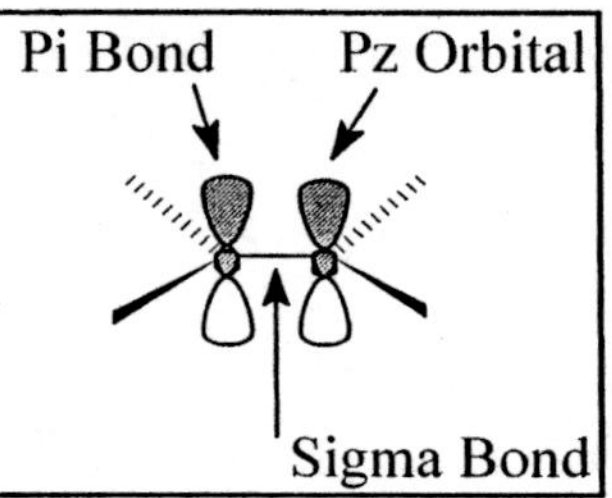

Pi Bond   Pz Orbital

Sigma Bond

The Baeyer Test (Aqueous Potassium Permanganate)

Potassium permanganate also undergoes an addition reaction with double bonds at room temperature, forming an intermediate, which reacts with water to form a glycol (dihydroxyalkane), manganese dioxide (a brown precipitate) and KOH, as illustrated below. The purple color of the permanganate is discharged (disappears) since it is completely converted to brown colored MnO_2. Sometimes it is mistakenly said that the solution turns brown. This is not true, the brown color is the precipitate.

$$H_2C=CH_2 \xrightarrow[H_2O]{KMnO_4} HO-CH_2-CH_2-OH + MnO_2 + OH^-$$

Reaction with H_2SO_4

Alkenes undergo an addition with sulfuric acid to form an alkyl hydrogen sulfate, which is soluble in the sulfuric acid. They can also form polymeric materials and tars, which may have a disagreeable odor and are brown to black in color.

$$H_2C=CH_2 \xrightarrow{H_2SO_4} H-\overset{\displaystyle H}{\underset{\displaystyle H}{C}}-\overset{\displaystyle OSO_2H}{\underset{\displaystyle H}{C}}-H \quad + \text{Polymers}$$

III. Alkynes

We will use propargyl alcohol (2-propyn-1-ol H-C≡C-CH$_2$OH) as the alkyne in the following tests. Alkynes with the triple bond at the end of the chain do undergo unique reactions which distinguish them from their relatives the alkenes. However, the reaction, replacement of the alkyne hydrogen with metals such as silver, yields metal acetylides (organometallic compounds) which explode when subjected to shock when dry. Thus, this is not a good experiment for a beginning lab. Alkynes contain the **triple bond, which reacts as though it were two double bonds** contained between the same two carbons. Remember that the triple bond is composed of a sigma bond and two pi bonds at right angles to one another (see the figure). These pi bonds undergo all of the same reactions as do the double bond systems: namely **addition of molecules** such as bromine, water, hydrogen and permanganate.

$$R-C\equiv C-R \xrightarrow{Br_2} \underset{R}{\overset{Br}{\diagdown}}C=C\underset{R}{\overset{Br}{\diagup}} \xrightarrow{Br_2} Br-\underset{\underset{R}{|}}{\overset{\overset{Br}{|}}{C}}-\underset{\underset{R}{|}}{\overset{\overset{Br}{|}}{C}}-Br \qquad R \quad \text{Triple Bond} \quad R$$

IV. Aromatic Compounds

Aromatic compounds, such as benzene, toluene and naphthalene are often drawn as though they contained double bonds. In fact, these molecules **do not have double bonds**, and a better way of representing them is with the aromatic "ring", indicating that there is a pi bond which is continuous around the ring. The following tests will indeed visually prove that aromatic compounds do not undergo the reactions of double bonds, but instead undergo a unique set of reactions called **"electrophilic aromatic substitution,"** represented by the sulfonation reaction below. Electrophilic aromatic substitution reactions involve replacing (substituting) hydrogens on the aromatic ring with electron deficient species called electrophiles (electron loving). Substituents on the ring, such as methyl or hydroxy, determine which ring hydrogens are replaced during the reaction. The methyl group donates some electron density to the 2, 4 and 6 (ortho and para) positions on the ring, making these positions more likely to react with the electron deficient electrophile, and also increasing the rate of the reaction, compared to a similar reaction with unsubstituted benzene. A more complete explanation will be found in your lecture text in the aromatic compounds chapter. We will also conduct a solubility test for toluene, to determine its relative polarity, and a flame test for toluene and, for comparison, cyclohexene and heptane. The flame test is often used to determine if a molecule contains unsaturation (a double bond or aromatic ring), since these materials often burn with a smokey/sooty flame.

V. Halogenated Hydrocarbons

We will prepare but not isolate brominated hydrocarbons through the free radical halogenation of alkanes reaction and the addition of bromine to alkenes. Chlorine will also undergo these reactions.. Fluorine is highly reactive, and can lead to uncontrolled reactions, including explosions. Iodine is essentially non reactive under these conditions. Chlorinated hydrocarbons are important industrial chemicals, since they are used to manufacture many products. We will perform two tests for chlorinated hydrocarbons. first, we will perform a solubility/density test. Chlorinated hydrocarbons are usually insoluble in water and also usually heavier (greater density) than water. We will perform this test with methylene chloride (dichloromethane), a common low boiling industrial and lab solvent. The second test will be a flame test. A rapid test to determine whether or not chlorine is present in a chemical is to burn a small sample in the presence of copper. If the flame turns green, the material contains chlorine. We will perform this test on a plastic, common PVC or "Vinyl," which contains a significant amount of chlorine.

Hazardous Materials

The following materials in this experiment can cause serious injury and require special care and possibly the use of gloves (use of goggles is required at all times in the lab).

1. *Hydrocarbons and acetone: a) remove oils from the skin causing cracking and itching, b) mild to severe discomfort if breathed in, c) cyclohexene has a disagreeable odor. All are highly flammable.*

2. *Sulfuric Acid: strong acid, causes severe burns within a few seconds.*

48

3. *Bromine in methylene chloride: bromine causes severe burn; CH_2Cl_2 is a potential carcinogen.*

4. *Aqueous (water solution) potassium permanganate: a strong oxidizer, can stain the skin.*

Protection against injury may require using gloves. However, the best protection is to be careful and to wash your hands on a regular basis while working with chemicals.

Glassware and Equipment

Nearly all reactions performed in the lab will be carried out in either a **4-inch, 5mL** *reaction tube or a 4 inch test tube.*

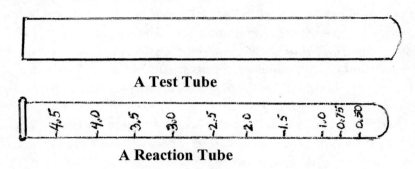

A Test Tube

A Reaction Tube

The reaction tube is narrower and thicker walled than a test tube. It also has volume markings on it. You will have red plastic caps for the tube and a cork or rubber stopper for the test tube, but you should not normally shake the tube up and down, since solvent and reagent could contact your skin when the cap is removed. You should therefore shake the tube using the finger flick method, which will be demonstrated by your instructor, or use a stir rod to mix the contents. If a reaction or solubility test uses water (i.e. aqueous solution), then the tube can be wet with water. For all other reactions, after disposing of the waste material, rinse with two x 1/2 milliliter portions of acetone (if the waste was an organic material) or water (if the waste was a water solution). If necessary, use a test tube brush and clean the tube with soap and water and some elbow grease. A plastic basin with detergent and test tube brushes is located on one of the benches. Then "dry" the tube by wiping the inside of the tube with rolled-up paper towel. You will use either a pump dispenser, Pasteur pipette or squeeze bottle to add reagents to the tube. The Pasteur pipettes and squeeze bottles are used to deliver material 1 drop at a time. If you should accidently drop a reaction tube or a test tube, don't worry about the cost, make sure that no chemicals splashed onto anyone's clothes or skin. Notify the instructor and follow directions for cleaning up the chemicals and glassware. Paper toweling contaminated with chemicals should be placed in the hood or rinsed well with water, never thrown directly into the trash. Sweep the broken glass up using a dust pan and brush, never use your fingers. The glass should then be disposed of directly into the glass disposal container.

Reagents for the Tests

Most of the reagents for the tests are all located in the hoods. Use the pump dispensers or dropping bulbs to obtain the chemicals. Whenever the experiment calls for the use of water, always use deionized water, not tap water (except for cleaning or when called for in an experiment). Use D.I. water to refill the wash bottles. Use tap water at your desk only to perform an initial water rinse of glassware. Follow this with a D.I. water rinse. Your instructor will show you how to use the pump and squeeze bottle dispensers. The pump dispensers are calibrated to dispense a set quantity of liquid. Use multiple strokes to obtain more than the set amount.

Disposal of Waste

*All organic waste must be placed in specially marked waste containers in the hood. Organic waste is poured into the proper waste container and then usually rinsed with 2 x 1/2 mL portions of acetone, which is deposited into an appropriate waste container. Most organic waste will be deposited in the **non-halogenated hydrocarbons** waste bottle. Any organics which contain chlorine or bromine are deposited in the **halogenated hydrocarbons** waste container. Water solutions of acids or inorganics are deposited in the **inorganics waste** container. Special wastes or contaminated glassware will be deposited in marked containers, as noted in the experiment or by your instructor. If you have any doubts about proper disposal of waste, **ask your instructor**. if a waste container is nearly full, please **inform your instructor**.*

Procedure

A note on measurements. Use your reaction tube, if you can, to accurately measure volumes. A measuring device is also provided for estimating volumes in a test tube. A measured dispenser is also used for many reagents, for convenience. Lastly, you can measure water and most organics by drop. 1mL of water is ~20 drops. 1mL of most organic liquids is ~30 drops. Try this with water and acetone, using a reaction tube to measure the volumes.

A) Sulfonation of Toluene

It is recommended that you or your partner perform the sulfonation of toluene at the beginning of the lab, to allow time for the reaction to go to completion.

1. Add 1 mL of concentrated sulfuric acid (**Caution**) to a clean, dry test tube, followed by 5 drops of toluene. Place the tube into a warm water bath (50°C, use a thermometer to measure the temperature of the water bath). *Every 5 minutes remove the tube and shake vigorously with the finger flick technique.*

2. After 30 minutes, or whenever the two layers have completely mixed, pour the contents into a 50mL beaker containing **(1) one** piece of ice. Set the tube in the test tube holder, after wiping the outside with a wet paper towel to remove any acid.

3. After the ice has melted, pour the solution back into the reaction tube and examine for solubility of the product.

4. Pour the contents back into the 50mL beaker and dispose in the non-halogenated hydrocarbons waste container. Rinse the reaction tube and beaker with acetone into the same hydrocarbons waste container and then clean them with tap water in your sink.

B) Solubility Tests

The solubility tests for heptane, cyclohexene, toluene and methylene chloride should all be performed at the same time. Use the same conditions for all tests.

1. Label 8 reaction tubes 1-8. Add 10 drops (1/2 mL) of blue-tinted water (in a dropping bottle on your desk) to tubes 1-4 and dispense 0.5 mL of ethanol into tubes 5-8.

2. Add 5 drops of heptane to reaction tubes 1 & 5, cyclohexene to reaction tubes 2 & 6, toluene to test tubes 3 & 7 and methylene chloride to tubes 4 and 8. Note whether or not the chemicals dissolve in the solvents. Shake the tube using the finger flick technique, as demonstrated by your instructor, let the contents settle for a few seconds, and again look for solubility. If the chemicals are soluble, there will be one layer. Note your observations on the data sheet at the end of the experiment.

3. Dispose of the water and alcohol mixtures in **tubes 1-3 and 5-7 in the non-halogenated organics waste container** in the hood and rinse the tube with two 1/2 mL portions of acetone into the **acetone waste container.** Pour the contents of tubes 4 and 8 into the **halogenated organics waste** container. Rinse the tubes with acetone.

C) Free Radical Halogenation

1. Add 15 drops of heptane (½ mL) to two clean and **dry** test tubes.

2. Add 5 drops of bromine dissolved in methylene chloride (CH_2Cl_2). The solution should be a dark yellow. If it isn't, add more Br_2/CH_2Cl_2.

3. Write your names on the two tubes (use label tape). Fold two pieces of blue litmus in half and drape over the tops of each test tube with one half inside each tube.

4. Place one tube under the **"black box."** Place the second tube under the **ultraviolet light.** After 10 minutes check both tubes. Work on one of the other experiments during this time. If a reaction has taken place, the yellow color will disappear and the blue litmus will turn pink, due to the liberation of HBr gas (acidic). Write down your observations and dispose of the liquids in the waste bottle labeled **"halogenated hydrocarbons."**

51

5. Rinse the tubes with two 1/2 mL portions of acetone and pour the rinses into the acetone waste bottle.

D) Addition Reaction with Br_2/CH_2Cl_2

1. Label three clean, dry test tubes 1-3. To tube 1 add 10 drops of cyclohexene, to tube 2 add 10 drops of toluene and to tube 3 add 10 drops of propargyl alcohol.

2. Add, a drop at a time, bromine in methylene chloride to each tube. Add 3-5 drops, noting whether or not there is an immediate disappearance of the bromine color as each drop is added (the propargyl alcohol may take a little longer to react).

3. Note your observations, dispose of the solutions in the **halogenated waste container** and rinse the tubes with acetone.

E) The Baeyer Test (Aqueous Potassium Permanganate)

The Baeyer tests for heptane, cyclohexene, toluene and propargyl alcohol should be done at the same time. Use the same conditions for all four tests.

1. Dispense 1 mL of acetone into four clean test tubes labeled 1-4. Add 5 drops of potassium permanganate solution to each tube and note the color of the solutions.

2. Add 2 drops of heptane to tube 1, cyclohexene to tube 2, toluene to tube 3 and propargyl alcohol to tube 4 and shake for a few seconds. Allow the mixture to settle and observe. If a positive reaction occurs, the purple color of the permanganate will disappear and a brown precipitate will appear. Write down your observations.

3. Dispose of the mixtures in the **non-halogenated organics waste** bottle, rinsing with1/2 mL acetone and then with water at your station.

F) Reaction with H_2SO_4
Use a 3 inch test tube for this reaction!

1. Dispense 1/2 mL of concentrated sulfuric acid (**Caution**) into a clean, dry 3-inch test tube (**do not use a reaction tube**) and add 5 drops of cyclohexene. Hold the tube only at the top. The reaction is very exothermic and the tube becomes very hot.

2. Shake the tube using the finger flick method (the sulfuric acid is very viscous) and observe whether there is a reaction.

3. Record your observations, take the tube over to the hood and carefully pour the contents into the non-halogenated organics waste container. Rinse the tube with two 0.5 mL portions of acetone into the waste container and then clean with water in your sink. Place the tube in the designated container in the hood

G) Flame Tests

1. Dispense 1/2 mL (about 15 drops) of heptane into a clean, dry test tube and then pour the liquid onto a clean watch glass in the hood. Immediately, before the material evaporates, light the fluid using a burning wood stick, which is lit in turn with a bunsen burner. Repeat the test using a clean watch glass with cyclohexene and then toluene. Compare the appearances of the flames and the amount of soot deposited on the watch glasses. Clean the watch glasses with paper toweling, which has been wet with 1 mL of acetone and return them to the hood immediately.

2. You will find a copper wire flame test probe attached to a cork handle in the hood for the chlorinated hydrocarbons flame test. Hold the tip of the wire in the hottest part of the flame until it is red hot. Immediately touch a piece of PVC polymer (in a watch glass next to the burner) with the hot wire tip until a little of the polymer has melted onto the wire. Return the wire to the flame and watch for a green color in the flame.

Properties and Reactions of Hydrocarbons
Data Sheet

Enter your data and observations as you perform the experiment. Explain your results, using reaction equations where necessary.

Br_2/CH_2Cl_2 Tests

Hydrocarbon	Results, Explanation, Reaction Equation
alkane (dark)	
alkane (UV light)	
alkene	
alkyne	
aromatic	

Baeyer ($KMnO_4$) Tests

Hydrocarbon	Results, Explanation, Reaction Equation
alkane	
alkene	
alkyne	
aromatic	

Sulfuric Acid Tests

Hydrocarbon	Results, Explanation, Reaction Equation
alkene	
aromatic	

Solubility Tests

Hydrocarbon	Results, Explanation	Results, Explanation
	Solubility in water	Solubility in Ethanol
alkane		
alkene		
aromatic		

Flame Tests

Hydrocarbon	Results, Explanation, Reaction Equation
alkane	
alkene	
aromatic	

Experiment 2
Questions

1. Why does the blue litmus paper turn pink during the free radical halogenation of an alkane?

2. Suppose an unknown liquid contained either p-xylene (1,4-dimethylbenzene), 1-hexene or benzene. How could you use chemical tests to tell whether the material contained 1-hexene and not the other two? *Hint! Draw these chemicals. What functional groups are contained in each?*

3. How could you use chemical tests to tell the above liquid was p-xylene and not the other two chemicals?

4. How could you use chemical tests to tell that the above liquid contained benzene and not the other two chemicals?

5. Which of the tests we performed would be positive when used for kerosene? Why? *Hint! Look up the definition of kerosene in your lecture text before answering this question.*

6. Gasoline is a complex mixture of alkanes and octane boosters, including aromatic compounds, alcohols or ethers. Alkenes are removed from gasoline because they cause the fuel to gum up on standing. How would you test for the presence of alkenes in a gasoline sample, using a chemical test?

7. Why would the flame test not be a good answer for question 6?

8. Write a balanced equation for the complete combustion of heptane.

Experiment 3
Dehydration of Cyclohexanol & Preparation of Cyclohexene

Record your observations and data on the report sheet at the end of this experiment.

Alkenes are commonly prepared in the laboratory and commercially by two related reactions: **dehydration** and **dehydrohalogenation**. Both reactions involve the removal of a molecule (water in the case of dehydration and HCl in the case of dehydrochlorination) from a hydroxy or halogen substituted alkane (see the reactions below).

The two reactions require quite different reaction conditions and reagents. Dehydration is accomplished under acidic conditions and dehydrohalogenation under basic conditions. In this experiment you will prepare cyclohexene by an acid catalyzed dehydration of cyclohexanol. The acid used will be dilute sulfuric acid. Concentrated sulfuric acid quickly dehydrates cyclohexanol, but leads to a good deal of polymeric side product as well. This is because the cyclohexene product reacts with the sulfuric acid as well to form polymer and other products (recall the reaction of cyclohexene with sulfuric acid in the hydrocarbons experiment). Phosphoric acid yields very little side product, but the reaction is slower. The mechanism for the reaction, shown below, involves protonation of the alcohol oxygen, followed by loss or water and then a proton.

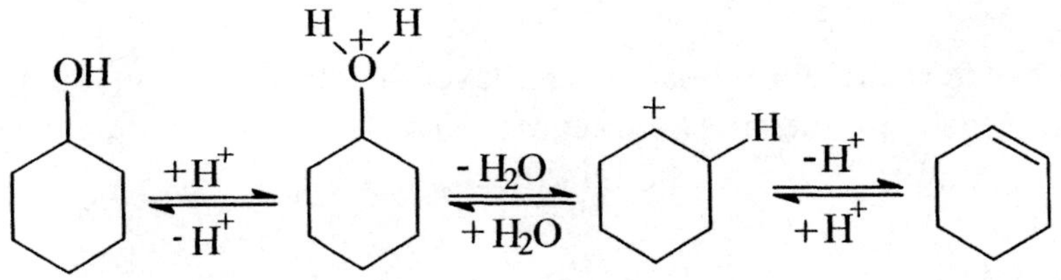

The acid is, in effect, a **catalyst** since H^+ is required to initiate the reaction but is regenerated in the process. The reaction, as noted, is actually reversible. An alkene, in the presence of acid and water, will add the water molecule to the double bond to form an alcohol. In our case, we wish to maximize the yield of alkene and must therefore force the reaction to the right, in the direction of formation of cyclohexene. This is done by removing the water and cyclohexene from the reaction flask as they are formed, by distilling (boiling) them off and collecting the liquids after cooling. Saturated sodium chloride solution is added to aid in separating the cyclohexene from the water. The cyclohexene is treated with calcium chloride to remove traces of water dissolved in it. The calcium chloride, in its anhydrous form, is referred to as a **drying agent** since it can remove small amounts of water dissolved in solvents by incorporating the water molecules into its crystal structure. Each $CaCl_2$ can incorporate 6 molecules of water to yield a crystal structure with the formula $CaCl_2 \cdot 6H_2O$.

Hazardous Materials

Cyclohexene and cyclohexanol are flammable and can cause skin and airway irritation. Sulfuric acid can cause severe skin burns and eye damage. Always immediately flush skin or eyes with water if you contact a chemical. Cyclohexene and the still bottoms have disagreeable odors. Handle carefully and always store and dispose of chemicals properly. Never pour any organic chemical into the sink. The heating mantle and distillation glassware becomes very hot and can cause burns.

Procedure

A. Dehydration of Cyclohexanol

Before you begin, your instructor will show you how to use the CRC handbook to obtain the physical properties (boiling points, density, etc.) for cyclohexanol and cyclohexene.

1. Take the clean, dry pear-shaped 10 mL distillation flask (**figure 1**) to the hood and dispense 0.7 mL of dilute sulfuric acid (0.3 mL H_2O/0.4 mL H_2SO_4) into the flask. Be careful not to get any acid into the distillation side arm.

2. Take the flask to the balance and tare the beaker on the balance. Place the flask into the beaker and weigh it to the nearest 0.01 g.

3. Take the flask to the hood and dispense 2.0 mL cyclohexanol into the flask, again being careful not to get any liquid into the distillation arm. Gently swirl the contents to mix the liquids.

4. Reweigh the flask to obtain the actual mass of cyclohexanol that you are using, to the nearest 0.1 g. Use this mass to calculate the number of moles of cyclohexanol you are using, given the mass of cyclohexanol is 100 g and the relationship:

$$\# \text{ moles} = \frac{\text{mass (g)}}{\text{MW (g/mole)}}$$

5. Add a boiling stone to the flask. Take the flask to your station and set up your distillation equipment as in **figure 1.** There should be an example set-up in the lab for you to examine. Make certain you clamp the flask at the indicated angle, so that distilled liquid will collect in the receiver reaction tube. Also, make certain that the thermometer is placed so that the top of the thermometer bulb is even with the bottom of the distillation side arm. Finally, be sure the heating mantle is plugged into the temperature controller (not into the wall plug) and the controller is off. Obtain your instructor's permission before proceeding further.

6. The flask should be immersed about half way in the sand and the receiver reaction tube immersed in ice/water. Turn the temperature controller to ~50%. It will take 10 minutes for the sand to heat up. You can control how hot the flask gets by pushing sand onto or away from the flask.

7. The liquid will begin to boil and eventually distill over into the receiver reaction tube. You may need to drape a piece of aluminum foil or paper toweling around the distillation flask to shield it from air drafts.

8. Continue distilling, noting the temperature range over which the liquid distills. You will obtain a mixture of water and cyclohexene. Eventually, the temperature will **begin to fall. You may also notice white smoke** in the flask. **You should have ~1 mL of cyclohexene (plus some water) in the receiver reaction tube at this point. Immediately** raise the flask (hold it with the clamp) from the sand bath. Turn off the temperature controller and allow the flask to cool.

9. While the flask is cooling, remove the receiver reaction tube and note the volume of cyclohexene (the upper layer). Add 2 drops of the blue dyed water to the tube to aid in distinguishing between the layers. Add ~ 1 mL of saturated sodium chloride solution to the mixture.

10. Cap the reaction tube with a red polyethylene cap and shake the contents for 10 seconds. Remove the cap and allow the layers to separate.

11. Hold, in one hand, an empty, dry reaction tube and the tube with your products next to each other in the test tube rack. Carefully use a pasture pipette (squeeze the air out of the bulb first) to remove the **top** cyclohexene layer into the pipette and then transfer it into the empty tube.

12. Add about 6-10 calcium chloride pellets to the tube with your product. Cap the tube with the red cap and **gently** shake the tube for (2) two minutes. Let the tube set for (1) one minute. Show it to your instructor if you are not sure that the product is dry.
    Your instructor may suggest adding more calcium chloride to finish the drying procedure.

13. Tare (weigh) a capped and labeled empty vial to the nearest 0.1g . Place the vial in the test tube rack next to the reaction tube. Place a special pipette (prepared for you) containing a plug of cotton (to remove $CaCl_2$ particles) into the vial.

14. Carefully insert **a clean** pipette to the **bottom** of the reaction tube, withdraw the dried product and transfer to the top of the filtering pipette in the vial. Allow the liquid to filter into the vial.

15. Cap the vial, re-weigh and write your names and the time and day of your lab on the label. The weight of the vial, mass of the liquid, etc. should be written down in the data section at the end of this experiment.

16. The bottle is placed in a box provided by your instructor to be stored in the refrigerator until next week, when you will obtain a gas chromatogram of the product.

Clean-up Procedure

Rinse out the distillation flask twice with 1.0 mL portions of acetone, disposing the waste in the non-halogenated hydrocarbons waste bottle. Leave the distillation flask in the labeled container in the hood. Pour the CaCl₂ pellets into the labeled beaker in the hood. Rinse the reaction tubes with two 1/2 mL portions of acetone into the acetone waste bottle and return them to your desk to dry. Wipe the thermometer with paper and leave it on your desk. Leave the pipette bulb on your desk. Place all pasture pipettes in the labeled container in the hood. Unplug the temperature controller.

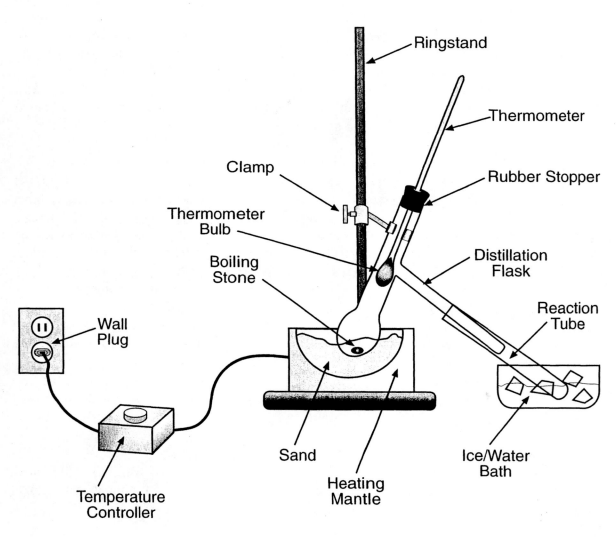

Figure 1.
Distillation Setup

Discussion on Calculations

An important part of the synthesis of chemicals is calculation of the proper amounts of reagents to be used in the reactions and the efficiency of the formation of products. To conduct these calculations it is first necessary to balance the chemical equation, in order to determine the stoichiometric or molar ratios of reagents and products for the reaction. For the present reaction, the acids used are catalysts and their concentrations theoretically do not change during the reaction. Even though this is not strictly true, we can indeed assume that their quantities are not critical for the determination of the yield of cyclohexene. Thus, the only reagent present which will determine how much product will be formed is the cyclohexanol itself and it is referred to as the **limiting reagent**. If we look at the overall chemical reaction, we see that the equation stoichiometry tells us that one molecule of cyclohexanol yields one molecule of cyclohexene. More importantly, one mole of cyclohexanol will yield one mole of cyclohexene. Thus, the **theoretical yield** of cyclohexene (in moles) will be the same as the number of moles of cyclohexanol that we started out with. How do we determine the number of moles of cyclohexanol that we started with? Remember that the **molecular weight** (molar mass in IUPAC nomenclature) is the sum of the atomic mass weights for the molecule. To an approximation this is 12g/mole x 6 = 72g/mole (for carbon) plus 16g/mole (for oxygen) plus 1 g/mole x 12 = 12g/mole (for hydrogen) = 100g/mole cyclohexanol. The actual numbers used will be determined by the number of significant figures that we can use in the calculations, based on the accuracy of the measurements in our experiment.

Now that we have the molecular weight for the cyclohexanol in g/mole, the actual number of moles of cyclohexanol used in the experiment can be determined by dividing the #g by the MW (g/g/mole = mole). Since this is also the number of moles of product and moles x MW = g (mole x g/mole = g), we can determine the theoretical number of grams of cyclohexene which should be obtained in the experiment (the theoretical yield). The molecular weight of cyclohexene is calculated using the same procedure outlined for cyclohexanol.

Finally, the **percentage yield** for the cyclohexene is calculated by dividing the actual yield in grams by the theoretical yield in grams x 100.

There is only one fly in the ointment: the product we obtained is not pure, it contains a small amount of starting material, as will be determined by GC. To be accurate, the theoretical and percentage yield must be corrected for the percent of cyclohanol in the product, as determined by GC. These calculations are outlined in the report for this experiment. The corrections require the data from the GC experiment, which you will obtain next.

For further examples of the use of balanced equations for calculating theoretical yields and percentage yields, do the appropriate homework problems in your lecture text.

Experiment 3
Dehydration of Cyclohexanol & Preparation of Cyclohexene Data Sheet

Observations and data (tares for vial, etc.) during the experiment.

1. Literature boiling point of cyclohexanol _____

2. Literature boiling point of cyclohexene _____

3. Actual boiling range of your cyclohexene product _____

4. Molecular Weight (Molar Mass) of cyclohexanol_____g/mole

5. Number of grams of cyclohexanol used _____g

6. Number of moles of cyclohexanol used _____mole

7. Theoretical number of moles of cyclohexene possible _____mole

8. Molecular Weight (Molar Mass) of cyclohexene _____g/mole

9. Theoretical Yield (number of grams) cyclohexene possible _____g

10. Actual Yield of cyclohexene (based on mass of liquid) _____g

11. Percentage Yield of cyclohexene (based on mass of liquid) _____%

The following calculations will be performed after you have obtained a gas chromatogram (GC).

12. Percent cyclohexene present in product based on GC results (cyclohexene area divided by cyclohexene plus cyclohexanol areas x 100)_____% (do not include water and methylene chloride areas obtained on the GC chromatogram)

13. Actual Yield of cyclohexene corrected by GC results (#12x #10)_____g

14. Actual Percentage Yield corrected by GC results _____%

15. Please comment on the purity and percentage yield for your product.

Experiment 3
Questions

1. You may notice that the chemicals in this experiment have rather distinctive odors. Compare the smell and appearance (at a distance) of cyclohexene, cyclohexanol, sulfuric acid to the smell and appearance of the still bottoms. Why are they different?

2. Would the yield of cyclohexene have increased if we had used pure sulfuric acid as the catalyst? Explain

3. What are some of the possible sources of loss of product in this synthesis?

4. If, instead of dehydration of cyclohexanol, we had prepared the product by dehydrohalogenation of bromocyclohexane, how much starting materials would have been needed to yield the *same theoretical yield* of product? Show calculations.

5. Why would the yield of product have been lower if we had not used the sodium chloride solution?

6. Why don't we include the GC area for the methylene chloride when we calculate the percent cyclohexene and cyclohexanol in your product?

Experiment 4
Chromatography

Chromatography refers to the various techniques used to separate and purify chemicals by passing their solutions in solvents or gases through a medium which retains each chemical to a different extent. There are two major types of chromatography: gas chromatography and liquid chromatography.

Gas Chromatography

In gas chromatography (GC) a small amount of liquid mixture or a mixture dissolved in a solvent is injected with a syringe into a hot vaporization port. The vaporized chemicals are then transported with a carrier gas (N_2, He, H_2 or Ar/Methane) onto a glass, metal or fused silica column containing a solid or high boiling liquid stationary phase which separates the chemicals in the mixture. As the separated chemicals exit the column, they pass through a detector, which then sends a signal to a recorder, integrator or computer, which displays a curve representing the amount of material verses time. A schematic for the GC is shown in **figure 1**. The detector we will be using is a "Hot Wire" or Thermal Conductivity Detector (TCD). It works on the principle that a wire with high resistance will carry a current which depends on the temperature of the wire, which in turn is dependent on the resistance of the wire and the cooling effect (thermal conductivity) of the carrier gas exiting from the chromatography column. Since the carrier gas (Helium for our system) is flowing at a constant rate (~30 cc/m), the wire will conduct a constant current. When a sample molecule hits the wire, it usually has a lower thermal conductivity than He, and the current decreases slightly. This is converted into an electric voltage output to the computer seen as the GC trace (figure 2). Since the temperature of the wire, and thus output voltage, are directly related to the number of molecules of sample hitting the wire, the area or height of each peak in the GC trace is directly related to the amount of material present (figures 3 and 4). If we injected a known amount of a pure chemical (a standard) into the GC and measured the area for the resulting peak, we could generate a standard curve to determine the absolute amount of that chemical in a mixture of chemicals. In our case, we will not inject known amounts of the standards, **we will simply compare that areas of the peaks to determine the percent of each component in the mixture**.

Separation of molecules by GC occurs because molecules vary in their structures, and thus physical properties. The property we will use to separate the chemicals is the boiling point. The column packing (a polymer powder) is coated with a thin layer of silicone oil. The individual molecules, carried by the He gas stream, dissolve in the oil. However, each molecule has a unique vapor pressure at a particular temperature and thus eventually exits from the oil and is carried through the column until it hits another particle. This process is repeated until the molecule exits from the column. Since vapor pressure and boiling point are related, **we are essentially separating chemicals according to their boiling points.** You will find that the components of your sample will separate according to the order of their boiling points, the lowest boiling first and the highest boiling last. Since the silicone oil is very non-polar, however, water, which is very polar, does not dissolve in the oil and is not retained on the column at all. Water (BP = 100°C), exits the column even before methylene chloride (BP=40°C). The time required for the injected chemical to exit the column is called the **retention** time. Since the vapor pressure for a chemical is constant at a constant temperature, if we inject a known pure chemical (a standard) into the GC, the retention time of the same chemical in a mixture will have the same retention time as the standard, as long as the operating conditions of the GC are identical. Thus, **we can identify a chemical in a mixture by its retention time**. Standards have been injected for you and are posted by the GC. Your retention time should be close to those of the standards, and the order that the chemicals come off the column will be the same. *One word of warning, methylene chloride will be used to clean your syringe. Therefore, there will be methylene chloride in your GC trace, but it is not one of your sample components.*

Thin Layer Chromatography (TLC)

In high performance liquid chromatography (HPLC), a small amount of a mixture dissolved in a solvent is injected with a syringe into a port containing a liquid being pumped at high pressures (1000 - 5000 lbs/in^2) onto a metal chromatography column containing a solid stationary phase which separates the chemicals. This process is analogous to GC, except that a liquid is used to carry the chemicals through the column. There are other differences also: instead of a silicone oil, the stationary phase is either silica gel (silicic acid, H_2SiO_3) or silica gel with octadecane (C18) bonded to it. Silica gel is very polar and bonds strongly to polar compounds. C18 is very nonpolar and bonds strongly to non polar compounds. Thus, these adsorbents separate chemicals based on their relative polarities. The detector for HPLC is usually an ultraviolet detector, which is sensitive to chemicals containing pi bonds and unshared electrons, such as aromatic rings, conjugated double bonds, aldehydes and ketones.

We will be using a simpler form of liquid chromatography, thin layer chromatography (TLC). TLC consists of a Mylar film about 1" x 4" coated with a thin layer of silica gel. The chemicals are dissolved in a liquid and a small spot applied at one end (**figure 6**). The plate is then immersed in a liquid, which is then allowed to migrate up the plate by capillary action. Since each chemical has a unique structure, they will bond to the silica gel to different extents. At the same time, each chemical will dissolve to a different extent in the solvent. Thus, the chemicals will migrate up the TLC plate, carried by the solvent, more or less, depending on its polarity. If standards are applied to the plate, they will migrate up a known distance, which can then be compared to the chemicals in a mixture, for their identification. The detector in TLC is our eye. The spots are either colored or made visible by shining UV light on them or by spraying them with a chemical which makes them visible. Therefore, this process of chromatographing the plate, followed by visualizing the spots is often referred to as "developing", similar to the developing of photographs.

I. Gas Chromatography

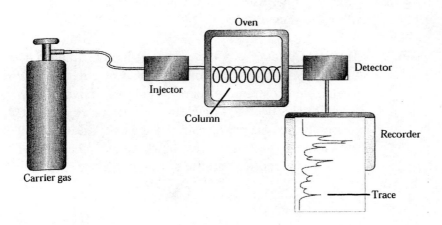

Fox-Whitesell, *Core Organic Chemistry*, 1997, Jones and Bartlett Publishers, Sudbury, MA www.jbpub.com. Reprinted with permission.

Figure 1. Schematic Diagram of a Gas Chromatograph

Gas Chromatography Terms

"Bleed" -   Small amounts of the liquid phase that are slowly eluted from the column when the maximum temperature for the column has been exceeded.

Carrier Gas -   Inert gas (usually helium, nitrogen, or sometimes methane/argon) that continually flows through the column.

Injection Port -   Entry port for sample. Sample must be injected (with a needle and syringe) past a rubber spectum which helps maintain a constant pressure of carrier gas (see figure).

Retention Time -   Period of time from injection required for the compound to pass through the gas chromatograph (see illustration).

Solid Support Material -   Inert high melting solid on which the stationary phase is coated. Example - firebrick. For a capillary column the stationery phase is coated on the wall of the column.

69

Stationary Phase -	Sometimes called the liquid phase. Usually a nonvolatile liquid or low melting solid. The portion of the packed column which interacts with the sample and affects separation. For capillary columns, the stationary phase is usually chemically bonded to the surface of the column to prevent bleed.
TCD -	(Thermal Conductivity Detector) - A hot wire placed in the gas stream at the column exit. The wire is heated by constant electrical voltage. When a steady stream of carrier gas passes over this wire, the rate at which it loses heat and its electrical resistance is constant. When the composition of the gas stream changes, the rate at which the wire loses heat changes, and therefore the resistance changes. This change in resistance is recorded on the strip recorder or computer as a peak.

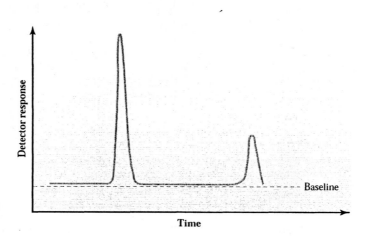

Fox-Whitesell, *Core Organic Chemistry*, 1997, Jones and Bartlett Publishers, Sudbury, MA www.jbpub.com. Reprinted with permission.

Figure 2. Retention Time

Factors Affecting Separation

1. Column Temperature - Generally low boiling compounds are eluted from the column faster than high boiling compounds because the low boiling compounds spend more time in the gas phase and are forced through the column quickly. **Increasing temperature decreases retention times.** If the column temperature is too low the high boiling compounds will be retained on the column. If the column temperature is too high all of the components will be swept from the column before they can separate. **A column temperature must be selected very carefully for good separation of the mixture.**

2. Carrier Gas Flow Rate - The carrier gas should not pass over the liquid phase so quickly that the mixture of compounds are swept away before they can equilibrate with the liquid phase. If this occurs, separation will not be optimum. If the flow is too slow the compounds will move through the column too slowly and will eventually be eluted from the column as broad bands. **Increasing carrier gas flow decreases retention times and sharpens peaks.** Carrier gas flows for packed columns are on the order of 30-40 cc/min. Capillary column flow rates range from 0.5 - 5 cc/m, depending on the diameter of the column.

3. Liquid Phase - The molecular weight, functional groups, and polarities of the compounds to be separated should be considered when choosing a liquid phase.

4. Column Length - The more similar the components of a mixture are in structure, functional group, and molecular weight, the greater the column length required for separation. **Increasing column length increases retention times**, but also improves separation of components.

Quantitative Analysis

The area under the gc peak is proportional to the amount of the compound that was eluted. Therefore, the molar percent composition can be calculated by comparing the relative peak areas. For a two peak mixture:

$$\% \ A \ = \ \frac{area \ of \ peak \ A}{area \ of \ peak \ A \ + \ area \ of \ peak \ B} \ X \ 100\%$$

$$\% \ B \ = \ \frac{area \ of \ peak \ B}{area \ of \ peak \ A \ + \ area \ of \ peak \ B} \ X \ 100\%$$

71

Hazardous Materials

Hydrocarbons are flammable and can cause irritation of skin and airways. Methylene chloride, benzene and azobenzene are potential carcinogens and should be handled with care. UV light will damage the eyes. Do not stare at the lamp even for a short period. The GC injector is set to 250°C and can cause burns. Cyclohexene has a particularly disagreeable odor and should be kept in a tightly stoppered container.

Procedure

Please note: You MUST clean the syringe with methylene chloride after its use. The cyclohexene polymerizes in the needle when placed in the hot injector. The methylene chloride will unplug the syringe for the next group.

1. Only analyze samples which have first been filtered to remove the drying agent which would otherwise plug the needle.

2. Rinse the 10 μL syringe by drawing 10 μL of methylene chloride rinse solvent into the syringe and then releasing the solvent from the syringe into a waste container. Repeat this process 6 times (if the syringe was not cleaned by the previous group). The remaining methylene chloride in the syringe (~1/2 μL) will not interfere with your analysis.

3. Pull the plunger of the syringe to the 2.0 μL mark, drawing air into the syringe.

4. With the needle of the syringe immersed in the cyclohexene sample, draw 1.0 μL of the sample into the syringe (the total volume air + cyclohexene = 3.0 μL).

5. Remove the needle from the sample, hold the syringe horizontally and pull the plunger out to the 5 μL position. You should now see a plug of approximately 1 μL of sample in the glass portion of the syringe, with air on either side.

6. Holding the syringe **vertically** with two hands (left hand holding the glass barrel and right hand holding the metal plunger), push the needle of the syringe needle all the way into the center of the injector A port (if you meet with resistance when inserting the needle, pull the needle out and reinsert it in the center of the injector A port).

7. Immediately, rapidly push the plunger in, forcing the sample and air into the injector port.

8. Immediately withdraw the needle and rinse six times with dichloromethane, injecting the rinse dichloromethane into the waste container.

9. Your partner must start the run on the computer at the same time as the sample is injected into the injector. Your instructor will show you how to do this.

10. The GC should detect three peaks (4 if any water is present, the first large peak will probably be dichloromethane solvent) during the analysis, requiring approximately four minutes. After the run is completed, the computer will integrate (determine the areas of) the peaks and print out a report. The system is then ready for the next injection.

11. Before you leave, write down all the chromatographic conditions (column type and length, temperatures, flow rates, etc.). Attach the chromatogram (or a photocopy) to your report.

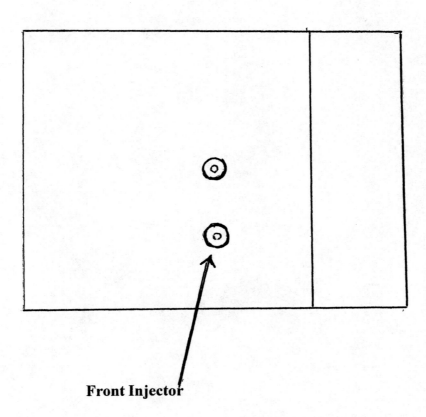

Figure 3. Top View of the Gas Chromatograph

73

Experiment 4
GC Analysis of Student Prep Cyclohexene
Data Sheet

1. GC Column_____
        (Length, Inner Diameter and Stationary Phase)

2. GC Column Temperature _____°C

3. Injector Temperature_____°C

4. Detector Temperature_____°C

5. Carrier Gas Flow Rate_____cc/min

6. Retention Time for Cyclohexene Standard _____min

7. Retention Time for Cyclohexanol Standard_____min

8. Retention Time for Sample Cyclohexene_____min

9. Retention Time for Sample Cyclohexanol_____min

10. Area for Cyclohexene in Sample _____ counts

11. Area for Cyclohexanol in Sample_____ counts

12. Total Cyclohexene + Cyclohexanol Area in Sample_____ counts

13. % Cyclohexene in Sample, based on areas in 10 and 11_____%

14. % Cyclohexanol in Sample, based on areas in 10 and 11 _____%

II. Photochemistry - Thin Layer Chromatography

troduction

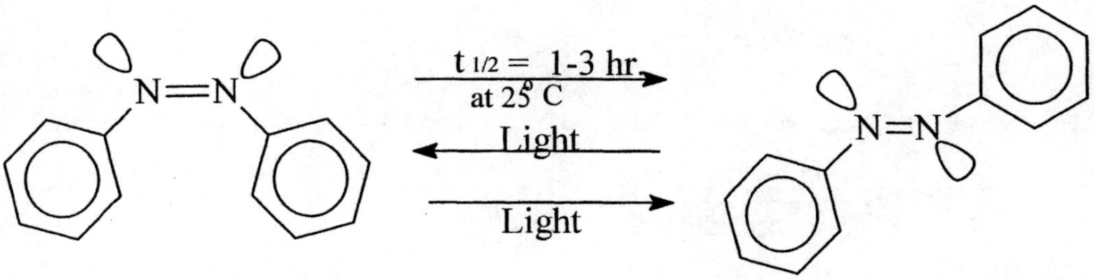

cis-azobenzene   trans-azobenzene

Figure 4.

Azobenzene belongs to a class of chemicals, called azo dyes, which form the largest group of
mmercial dyes. In addition to being used to dye cloth and color foods, some of the azodyes are also
ed in the chemistry lab as acid-base indicators. The azobenzenes contain a nitrogen-nitrogen double
nd and thus there exists the possibility of cis-trans isomerism. However, the trans form is usually
ore stable, since there s a good deal of steric interaction between the benzene rings in the cis form.
fact, a solution of cis-azobenzene will slowly convert in a few hours to trans-azobenzene at room
mperature - a process which requires breaking the π bond, rotation about the N-N single bond, and
formation of the π bond.

However, if one places a solution of pure trans-azobenzene directly in strong sunlight, some
nversion to the cis-azobenzene isomer occurs. Since heat converts cis- to trans-azobenzene, it must
the light which is responsible for the formation of the cis isomer. Sunlight, as it reaches the
ound, is principally visible light but also contains infrared light (responsible for the warmth we feel)
d ultraviolet light (responsible for a suntan or burn). The wavelength ranges and energies at the
ansitions between these three areas of the light spectrum are shown below. Notice that, as the
avelength becomes shorter, the energy increases.

We will perform an experiment to determine which part of the spectrum is responsible for the
terconversion of trans- to cis-azobenzene and the approximate energy necessary to interconvert the
o isomers.

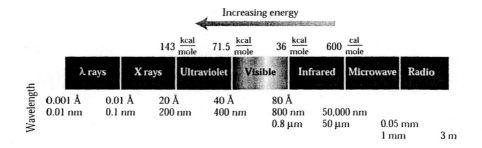

Fox-Whitesell, *Core Organic Chemistry*, 1997, Jones and Bartlett Publishers, Sudbury, MA www.jbpub.com. Reprinted with permission.

$$\text{Energy} = \frac{2.857 \times 10^{-3}}{\text{wavelength}} \text{ k cal} \cdot \text{cm/mole}$$

Figure 5.

The progress of the reaction will be followed by means of a technique called thin layer chromatography (TLC). This method is widely used in industry for the separation, identification, and quantification of mixtures of chemicals. Chromatography, in general, affects the separation of chemicals based upon differences in a physical property such as polarity. The sample is dissolved in a liquid which then flows through a stationary phase. The individual components of the mixture are held up in their passage through the stationary phase to differing extents, thus separating them. In this experiment the stationary phase is a thin layer of silica gel coated onto the surface of a "Mylar" plastic sheet.

In practice, four spots of a benzene solution of azobenzene will be placed on the TLC plate near the bottom. The first spot will be exposed to ultraviolet (UV) light while the others are covered. The second spot will be exposed to infrared (IR) light and the third spot to visible light. The fourth spot will not be exposed to light, but will be used as a trans-azobenzene standard. The TLC plate will then be placed in a container with a solvent mixture touching the bottom of the sheet. The solvent will travel upwards by capillary action (this is called developing) and will carry the trans- and/or cis-azobenzenes upward with the solvent. However, since the silica gel is polar and "holds on" to polar compounds more tightly than non-polar compounds, the distance the two compounds will travel up the plate will depend on their individual polarities. Since cis-azobenzene has more electron density (the unshared pair of electrons on the nitrogens) on one side of the molecule than trans-azobenzene, which has its electron density symmetrically distributed on both sides of the molecule, cis-azobenzene will be considerably more polar than trans-azobenzene. Thus we would expect cis-azobenzene to bind more tightly to the silica gel and not to move up the TLC plate as far as trans-azobenzene.

The distance that a chemical migrates up a TLC plate, divided by the distance the solvent front travels, is defined as the R_f value. Under a constant set of conditions (solvent, stationary phase, spot concentration, temperature, etc.) the R_f value will be a physical constant for a chemical much as the melting point and color are. Thus, an unknown chemical or component of a mixture can be identified if it has the same R_f as a known standard.

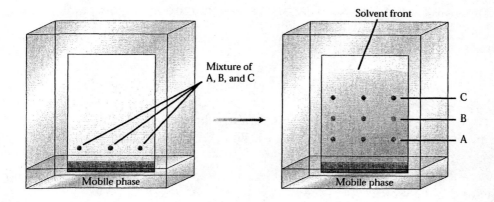

Fox-Whitesell, *Core Organic Chemistry*, 1997, Jones and Bartlett Publishers, Sudbury, MA www.jbpub.com. Reprinted with permission.

Figure 6. TLC Chromatography

PROCEDURE

A. Photochemistry

1.) Obtain a 1" x 3" TLC sheet, protect the coated side (the dull side) from mechanical injury, fingerprints, etc.

2.) Using a ***lead pencil***, place four ***faint*** pencil marks 1/2" from the bottom of the sheet as illustrated. Do **not** gouge a hole in the silica gel.

3.) Dip a microcapillary tube in the trans-azobenzene solution, gently touch the end to one of the spots and immediately remove it from contact. Repeat this procedure at each spot two more times or until you have enough material on the plate to easily see, but keep the spots ***two millimeters*** or less in diameter.

4.) Mask the second, third and fourth spots with foil and paper clips.

5.) Expose the first spot to IR light for 5 minutes.

6.) Remove the foil from the second spot and over the first one. Expose the second spot to UV light for 5 minutes.

7.) Remove the foil from the third spot and cover the second one. Expose the third spot to visible light for 5 minutes.

8.) While exposing the TLC sheet to the light, set up the chromatography tank as outlined below.

B. Chromatography

1.) Place 1/4" of solvent (3/1 mixture of cyclohexane/toluene) in a screw top bottle. Make sure there is a piece of filter paper in the bottle, touching the solvent. Tilt the bottle to soak the filter paper (the paper acts as a wick, saturating the atmosphere in the bottle with solvent). Place the lid on the bottle and let the bottle sit until the TLC plate is ready for development.

2.) Remove the foil masks and place the TLC sheet, spots at the bottom, into the bottle, tilting the top of the sheet *away* from the filter paper.

3.) Replace the cap *immediately* and let the solvent migrate up the plate until the solvent front is about 1/4" away from the top. Remove the sheet, recap the jar and use a pencil to lightly mark the solvent front on the TLC plate and observe the results. Circle the spots lightly with the pencil, since they will diffuse and become less visible over 1-2 weeks. *The cis-azobenzene will be a faint yellow spot near the bottom of the plate, close to the origin*.

4.) Return the chromatography jar, brass shims, paperclips and rulers when you are finished with them.

5.) Staple the TLC plate to your data sheet (the partner should draw an exact copy). Determine the distance (in centimeters) from the point of origin of each spot to the solvent front. Also determine the distance from the point of origin to the center of each spot. Remember that the 4th spot is the pure trans-azobenzene standard.
Calculate the R_f values and answer the questions.

Experiment 4
Photochemistry-Thin Layer Chromatography
Data Sheet

1. Place your TLC plate or drawing here. Indicate which spots were UV, IR or visible light exposed. Label the developed spots as cis- or trans-azobenzene.

2. Calculate the R_f values for each spot. Place the R_f values next to each spot.

3. What are your conclusions regarding the;

   (a) UV light effect on isomerization

   (b) IR light effect on isomerization

   (c) Visible light effect on isomerization?

4. What is the approximate minimum energy needed to isomerize the double bond.

5. Based upon the intensities of the spots, estimate the percent conversion of trans to cis-azobenzene.

6. Why are many beverages, chemicals, medicines, etc. stored in amber colored bottles.

Experiment 5
The Properties and Reactions of Alcohols, Phenols and Amines

I. Alcohols and Phenols

The alcohols and phenols are closely related functional groups. Alcohols are symbolized R-OH and phenols are symbolized Ar-OH. The "R" is any non aromatic carbon group while "Ar" represents any aromatic group. Typical alcohols and phenols encountered in the lab and industry are shown in **figure 1.**

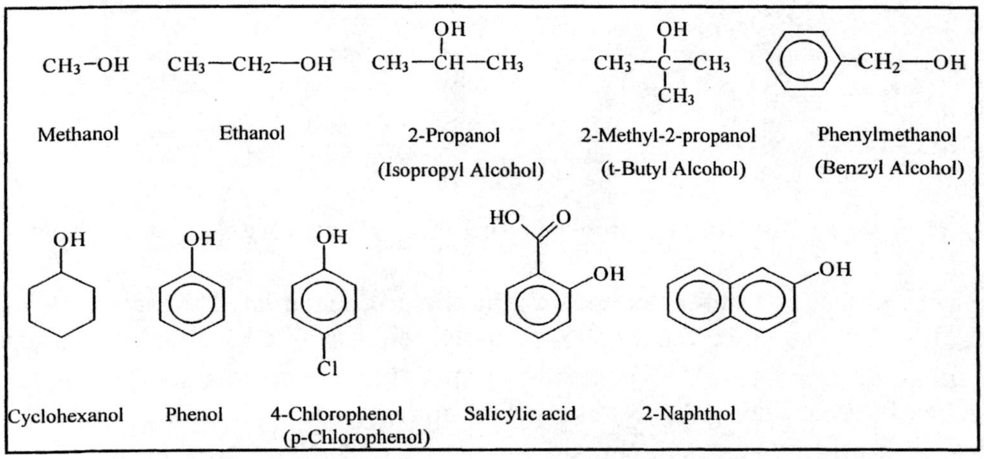

Figure 1. Common Alcohols and Phenols

The alcohols and phenols are considered separate functional groups since the aromatic ring alters the chemistry of the hydroxy group, giving phenols some properties and reactions quite different from those of alcohols. The following experiments illustrate the properties and reactions of the alcohols and phenols.

1. Solubility Properties

Most lower molecular weight alcohols are liquids. Most phenols are solids. Two of the phenols we will be testing, phenol and 4-chlorophenol (p-chlorophenol), have been "liquified" by adding a small amount of water. They are thus very concentrated aqueous solutions and are particularly hazardous if they contact the skin. The hydroxy group dominates the properties of alcohols and phenols. Lower molecular weight phenols and alcohols are soluble in water, since they form strong hydrogen bonds to water, as indicated in **figure 2**. Additional carbons and other functional groups, however, will change the solubility characteristics. Additional carbon groups lead to decreased water solubility as the nonpolar character of the molecule increases. On the other hand, even low molecular weight alcohols, such as methanol, are generally soluble in many organic solvents such as hexane.

Alcohols are neutral when dissolved in water. Phenols, on the other hand, range from very weak to very strong acids when dissolved in water, since the aromatic ring stabilizes the resulting anion, as shown in **figure 2**. Electron withdrawing substituents, such as nitro or chloro, at the ortho and para (2, 4, 6) positions increase the acidity and electron releasing groups decrease the acidity of phenols. Thus, a compound which forms a weak to strongly acidic solution, or dissolves in 5% sodium hydroxide solution may be a phenol.

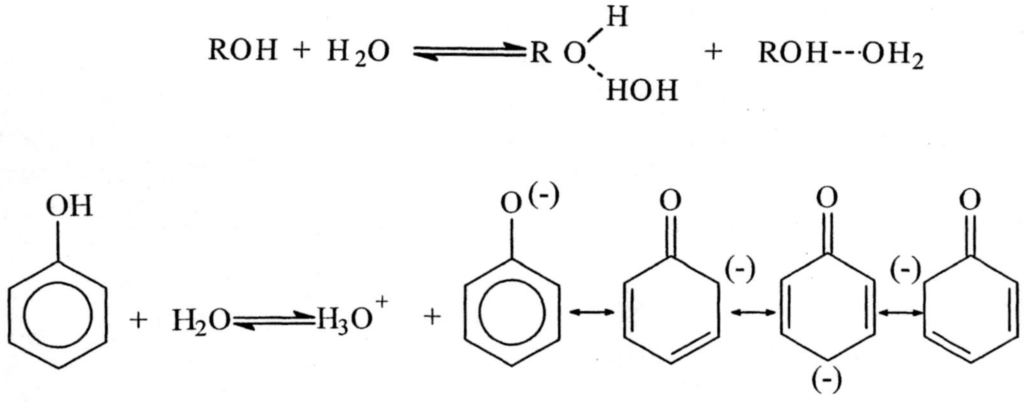

$$ROH + H_2O \rightleftharpoons R\text{-}O\underset{HOH}{\overset{H}{\diagdown}} + ROH\text{-}\text{-}OH_2$$

Figure 2. Hydrogen Bonding of Alcohols and Phenols and Acidity of Phenols

A common classification test for alcohols is solubility in concentrated phosphoric acid. Actually, this is not a solubility test, since the alcohol actually behaves as a base and is protonated by the phosphoric acid, as seen in **figure 3**. The resulting salt is soluble in the phosphoric acid. Often, the white salt can be observed during the test as a gelatinous ring where the two layers come in contact. If a stir rod is used to mix the layers, the salt does eventually dissolve. Alcohols are further classified by their reactions with sulfuric acid. The reaction can be quite vigorous and cause the alcohol solution to splatter, so we will not do this test here.

$$ROH + H_3PO_4 \rightleftharpoons ROH_2^{(+)} + H_2PO_4^{(-)}$$

Figure 3. Reaction of Alcohols with Phosphoric Acid to form a Soluble Salt

2. Reactions of Alcohols

Alcohols undergo four general reactions, illustrated in **figure 4**, (1) elimination of water or dehydration, covered in the cyclohexanol experiment, (2) substitution of the OH group, (3) substitution of the H on the oxygen, and (4) oxidation. The Lucas test for alcohols, illustrated in **figure 5**, is used to distinguish between primary (1°), secondary (2°) and tertiary (3°) alcohols. The Lucas reagent is a mixture of $ZnCl_2$, a strong Lewis acid and HCl, a strong Lewis/Bronsted acid. The reagent dissolves the alcohol and removes the OH group from the carbon, leaving behind a carbocation. Carbocation formation requires energy, and the more energy required to form the cation, the slower it forms. Tertiary, benzylic and allylic cabocations form very quickly, secondary carbocations form more slowly and primary carbocations form very slowly under the reaction conditions. When the carbocation forms, it immediately reacts with chloride ion and an insoluble cloroalkane is formed. Thus, immediate formation of a cloudy or oily looking mixture is an indication that the alcohol is 3°, allylic or benzylic. Formation of a cloudy mixture after about 5 minutes indicates the alcohol is 2°. Primary alcohols do not react at all within the time scale of the lab experiment.

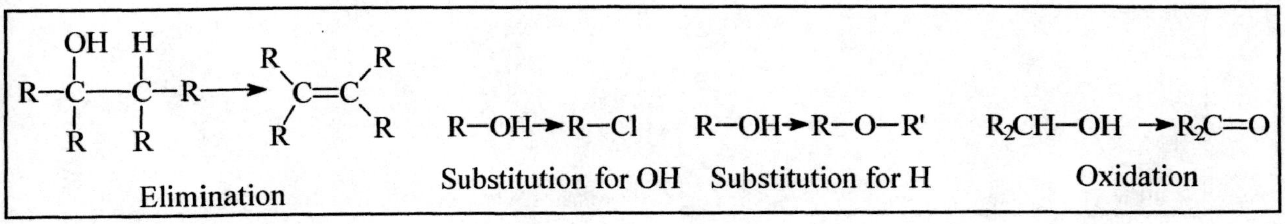

Figure 4. Four General Reactions of Alcohols

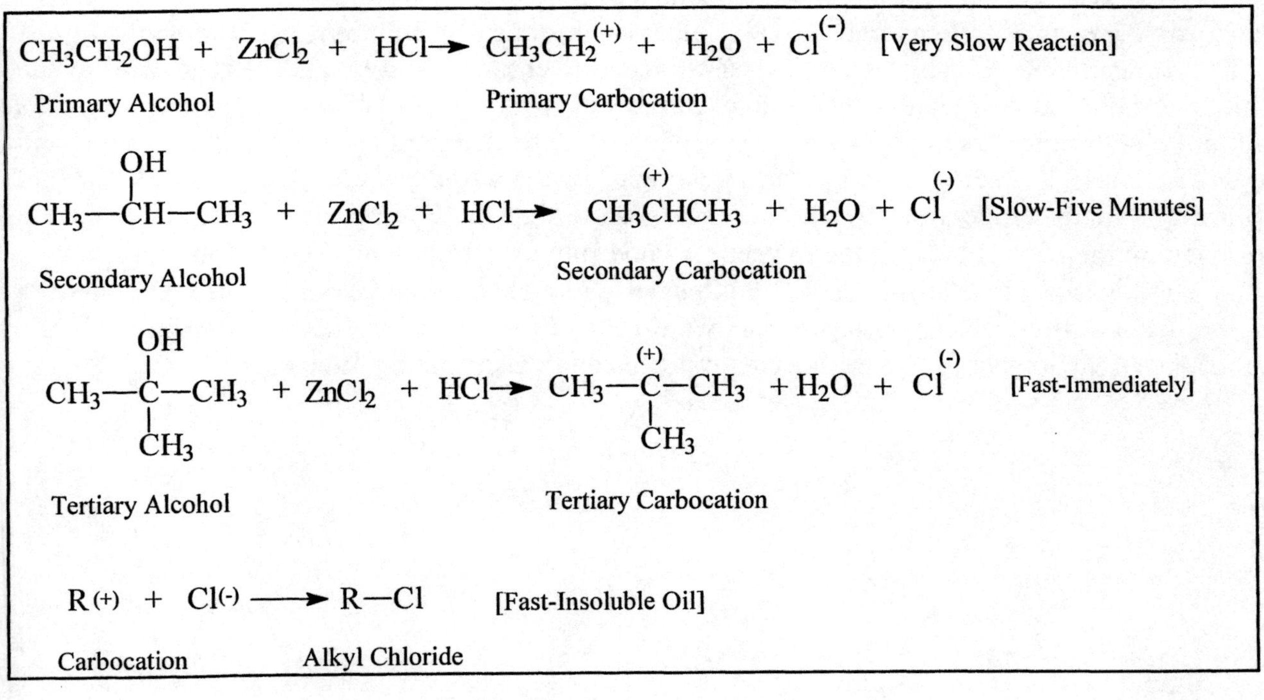

Figure 5. The Lucas Test for Alcohols

Substitution of the alcohol hydrogen is illustrated in **figure 6**. These reactions are not only used to classify and identify alcohols, but more importantly used to synthesize esters, ethers and urethanes, as well as the corresponding polymers. Due to lack of time, we will not perform these reactions here.

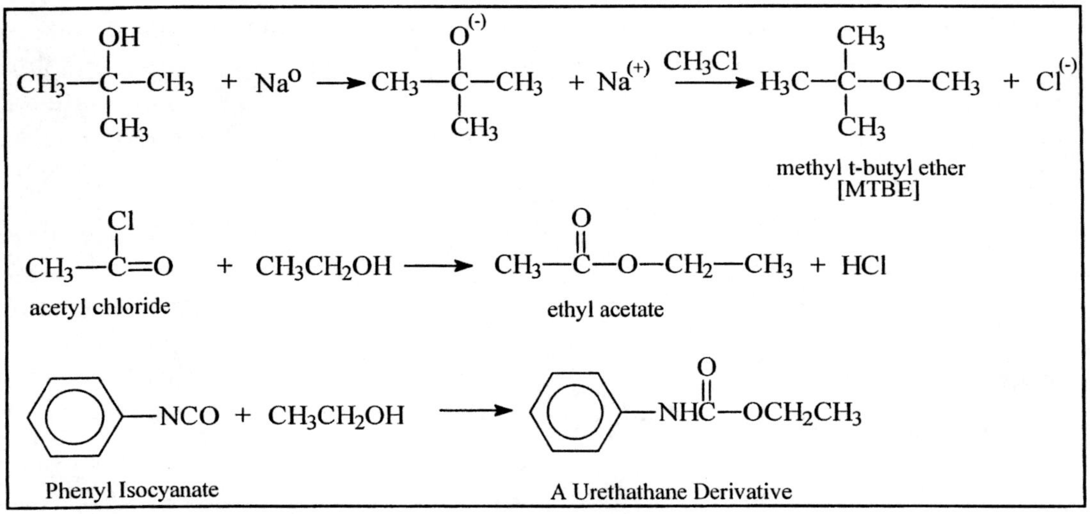

Figure 6. Substitution of the Alcohol Hydrogen

Oxidation of alcohols is used synthetically to prepare aldehydes, ketones and carboxylic acids. The reaction is also useful for classifying alcohols as 1°, 2° or 3°. Primary (1°) alcohols are very quickly oxidized at room temperature to aldehydes by chromic acid solution, as illustrated in **figure 7**. The corresponding aldehydes are further oxidized by chromic acid to a carboxylic acid. If the solution is then heated, low molecular weight carboxylic acids boil out of the solution and the vapors will turn blue litmus paper red. Secondary alcohols (2°) are also oxidized quickly by chromic acid to yield ketones. The ketones are not further oxidized by the reagent. Tertiary alcohols (3°) are not oxidized at room temperature by chromic acid. Note that the tertiary alcohol does not have a hydrogen on the carbon bearing the oxygen. A good rule of thumb is that **oxidation is occurring when an organic compound loses 2 hydrogens, gains an oxygen, loses an electron or a combination of these changes. Reduction would then be the opposite situation.** Apply this rule to the above reactions to note what is being oxidized and what is being reduced.

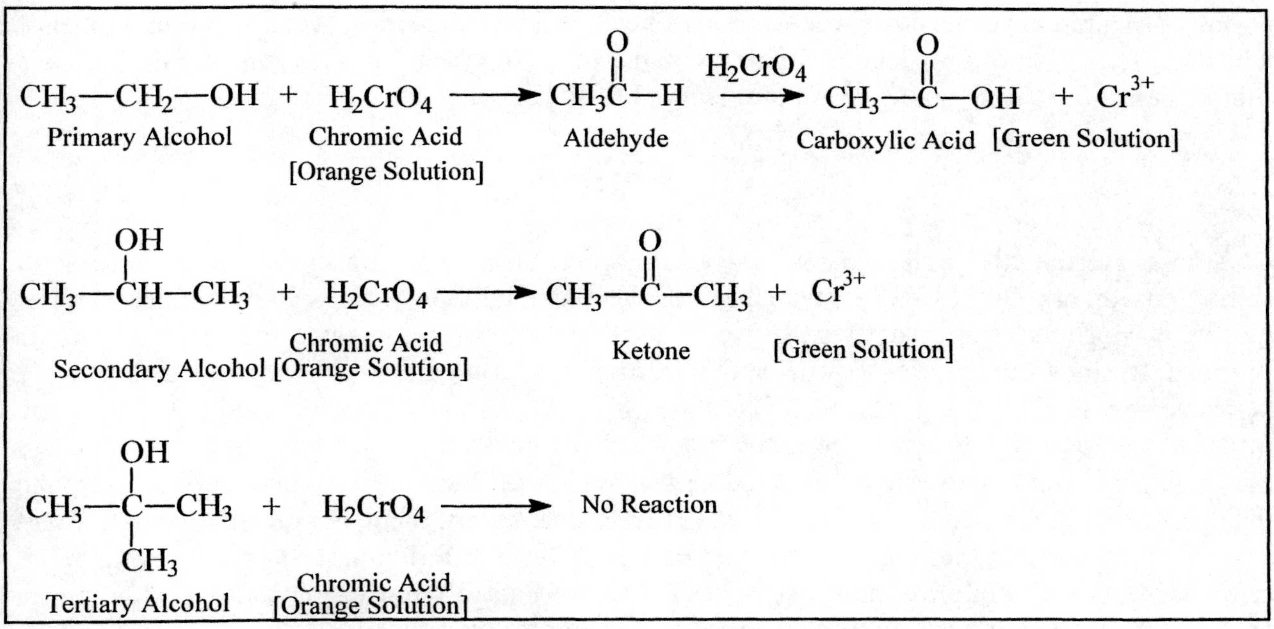

Figure 7. Chromic Acid Oxidation of Alcohols

3. Reactions of Phenols

Phenols will undergo many of the same reactions as alcohols, although reaction conditions are often different. We will use two additional reactions that phenols undergo and alcohols do not to illustrate classification reactions for these materials. Phenols are hydroxyaromatic compounds. Thus, phenols undergo the electrophilic aromatic substitution reactions of aromatic compounds. The hydroxy group is strongly electron donating to the aromatic ring and, as a result, increase the rate of the reaction. Electrophilic aromatic bromination of a phenol will occur even without adding a Lewis acid catalyst such as $FeCl_3$. Thus, a positive indication of a phenol is an immediate reaction of the material with a bromine/water solution to form an insoluble bromo product, as illustrated in **figure 8**. Bromine water will react with some other aromatic compounds, such as anilines and phenyl ethers but the acid/base tests and IR spectra should generally rule out other possibilities.

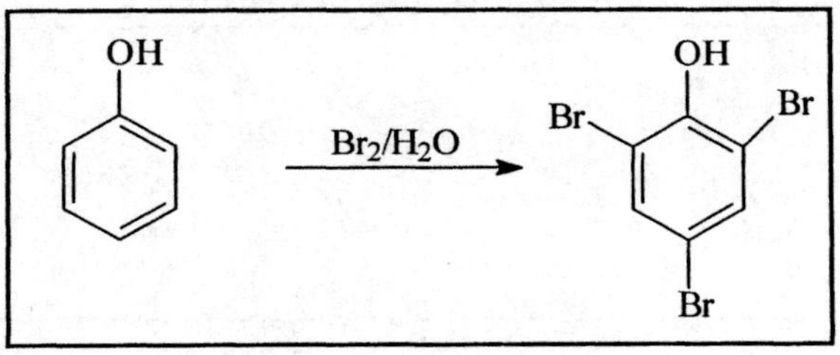

Figure 8. Bromination of Phenols

Many phenols also strongly bond to ferric chloride ($FeCl_3$) to form a colored complex. Thus, formation of a red to purple color when a drop of $FeCl_3$ is added to a dilute water solution of phenol indicates that a phenol is present. There are some other functional groups which form colored solutions also, but this is not usually a problem in identifying phenols.

II. Amines

Amines are the most basic organic functional group. Most amines react with water to form an ammonium ion and hydroxide ion, which makes the water solution basic. A few amines, such as pyrrole, do not react significantly with water but do dissolve when reacted with strongly acidic solutions. Amines are classified as primary (1°), secondary (2°) or tertiary (3°), depending on whether they have one (RNH_2 1°), two (R_2NH, 2°) or three (R_3N, 3°) carbon groups attached to the nitrogen. In addition, amines which have an aromatic ring attached directly to the nitrogen, such as aniline, are usually referred to as **aromatic amines** and amines which have the nitrogen as part of an aromatic ring system, such as pyridine or pyrrole, are referred to as **heterocyclic aromatic amines**. These terms are used because these different classes of amines have differing base strengths and undergo different reactions. While we cannot explore all of the various differences in reactions of amines in this lab period, we will perform some simple tests to classify the amines.

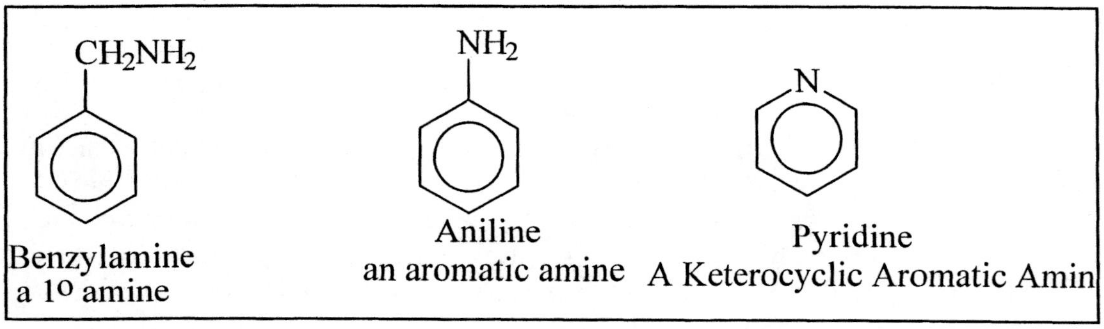

A. The Basic Nature of Nitrogen

All of the reactions of amines depend on the ability of the nitrogen atom to share the election pair on the atom with an electron deficient atom or ion, such as a proton, H^+. This reaction is illustrated below with reactions with water and hydrochloric acid in **figure 8.**

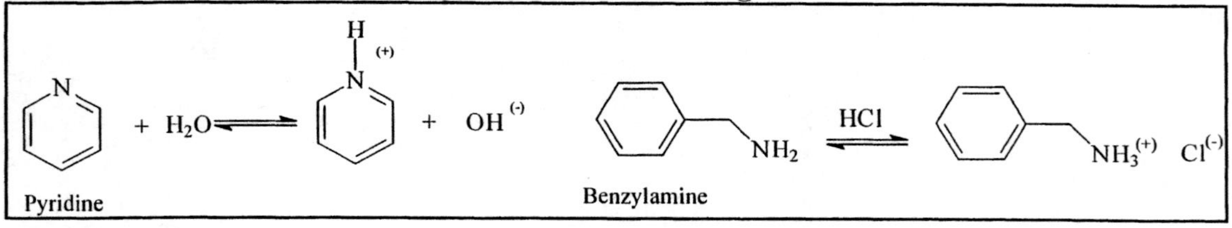

Figure 8. Reaction of Amines with Acids

The electron rich, Lewis base nitrogen can also react with Lewis acids other than H^+, including the carbonyl carbon of an acid chloride, ketone or aldehyde (the nitrogen of 2,4-DNP, used in the aldehydes and ketones experiment is a member of the amine family).

Hazardous Materials

Precautions must be exercised when handling phenols, especially liquid phenols and aqueous solutions. Phenol solutions can range from near neutral to very acidic. At the same time, phenols tend to be anaesthetics, local pain killers. Thus, if a phenol contacts the skin, it will numb the nerve while at the same time severely burning the skin. It is possible to obtain a third degree chemical burn within a few minutes of contact with a concentrated aqueous solution of phenol (liquified phenol) and not realize the damage until noticing the bleached white skin resulting from the contact. Wear gloves when handling glassware containing phenols and wash your hands and glassware frequently and thoroughly with large amounts of water. Most phenols and some alcohols have obnoxious odors. Do not discard any organic residue into the sink. If any does spill, wipe it up (use gloves) and rinse the paper thoroughly with water. We will also be using hydrochloric acid, sulfuric acid, chromic acid, sodium hydroxide and bromine. All will cause skin burns. Chromic acid causes cancer and must be treated as indicated in the experiment to convert it to a less hazardous Cr^{3+} form before discarding in the inorganic waste container.

Disposal of Hazardous Waste

Follow the general method for disposal of hazardous waste, pouring test tube contents into the non-chlorinated organics or chlorinated organics waste bottles, as indicated for each test. Rinse each tube with an additional 2-3 portions of ½-1 mL of acetone. The chromic acid waste will be poured into a special bottle, however, and rinsed with water into the same bottle, before rinsing with acetone.

Procedure

A. Solubility of Alcohols and Phenols in Water

1) Add 1 mL of **blue dyed deionized water** to each of 6 reaction tubes. Add 1 mL of clear water to a 7th test tube. Label the tubes and add 2 drops of the following alcohols separately to the first four tubes: ethanol, tert-butyl alcohol (2-methyl-2-propanol), cyclohexanol, and benzyl alcohol (phenylmethanol). To the fifth and sixth tubes add separately 2 drops of phenol and p-chlorophenol (4-chlorophenol) solutions, respectively. To the last tube add separately a few crystals of salicylic acid (2-hydroxybenzoic acid), using the small spatula provided.

2) Note whether or not the liquids and solids dissolve immediately, after mixing with the stir rod. If they do not immediately dissolve, place the tubes into a hot (~50°C) water bath for five minutes and check for solubility after again mixing with the stir rod.

3) Use the stir rod to remove a drop of solution from the ethanol and phenol solutions and test for acidity on two 1/4" strips of pH paper (not litmus paper) provided by your instructor. Compare these results to the pH results for a drop of *tap water*.

4) Pour the alcohol/phenol solutions **(tubes 1-5)** in the *organics waste container*, rinsing with two to three 1/2 mL portions of acetone.

5) Tube **6** will be tested with bromine water and tube **7** tested with ferric chloride (see below).

B. Reaction of Phenols with Bromine Water

1) To the water solution containing 4-chlorophenol **(tube 6)**, add 1 mL of bromine water solution (1 dropper bulb full). Note whether a precipitate is formed, whether or not it dissolves and its color.
2) Pour the mixture into the *halogenated organics waste container* and then rinse the tube with acetone.

C. Reaction of Phenols with Ferric Chloride

1) To tube 7, containing salicylic acid, add 1 drop of ferric chloride solution. Does the solution become colored? Remember, $FeCl_3$ solution is yellow!

2) Pour the solution in the *organics waste container* and then rinse the tube with acetone.

D. Solubility of Phenols in Sodium Hydroxide Solution

1) Add a few crystals of 2-naphthol to a test tube, followed by 1/2 mL water. Does the phenol dissolve? Now add 1/2 mL of 5% sodium hydroxide to the reaction tube and again check for solubility.

2) Pour the contents of the tube in the *non-halogenated waste container* and then rinse the tube with acetone.

E. Reaction of Alcohols with the Lucas Reagent

1) Add 1 mL of the Lucas reagent to 3 test tubes. Add, separately, to the marked tubes 4 drops of the following alcohols: ethanol, cyclohexanol and tert-butyl alcohol. Record the length of time it takes for the solution to become cloudy or to form an oily layer.

2) If the reaction does not take place within 2 minutes, place the tube into a warm (~40°C) water bath for 5 minutes, cool the tubes and observe.

3) Pour the solutions in the *halogenated organics waste container*, rinse the tube twice with water into the *halogenated organics waste container* and then rinse the tube with acetone.

F. Oxidation of Alcohols

The chromic acid solution has been prepared for you by adding sulfuric acid (H_2SO_4) to $K_2Cr_2O_7$. Thus, it is a strong acid and a powerful oxidant. Use proper precautions when using this reagent.

1) Place 0.5 mL of chromic acid solution into each of three 10 mL Erlenmeyer flasks. Add 4 drops of the following alcohols, separately, to the three labeled flasks: ethanol, cyclohexanol and tert-butyl alcohol. Note whether or not a color change takes place. A positive reaction is indicated by a change in color from orange (Cr^{6+}) to dark green (Cr^{3+}).

2) Wet two pieces of blue litmus paper with D.I. water (do not touch the middle of the paper with your hand) and drape it over the top of the flasks which turned color. Place these flasks on the hot plate and watch carefully. Bring the solution just to a simmer (do not boil) and immediately remove them from the hot plate. Watch the paper. Does it turn color?

3) When finished, ***add 1 mL of ethanol to all three flasks*** to reduce remaining Cr^{+6} to Cr^{+3} and pour the material into the ***inorganics waste container***, and rinse the tube twice into the same container. Then, rinse the tube with acetone.

G. Basic Properties of Amines

1) Into three separate test tubes containing 1 mL of blue dyed water, add (2) two drops of aniline, benzylamine, and pyridine respectively (what class does each amine belong to?). Mix the contents using the finger flick method. Check for solubilities and record your results. **If necessary**, heat the samples in a hot water bath (~50°C) for five minutes to aid solution.

2) **Using the stirring rod, remove a drop of the water layer** (the amine, if insoluble in water, may float on top if its density is less than 1) and check the pH using a 1/4 inch piece of Hydrion pH paper. Compare to the pH of 1 drop of tap (not deionized) water. Is there a difference in base strengths for the various classes of amines?

3) To each test tube add concentrated HCl, dropwise, until the solutions turn blue litmus distinctly red. What is the white "smoke" seen during the addition of the acid? Are the amines now all soluble? Dispose of the pyridine solutions in the non-halogenated organics waste container in the hood.

4) Rinse the tubes with two-three 1/2 mL portions of acetone into the then rinse the tube with acetone.

Experiment 5
The Properties and Reactions of Alcohols, Phenols and Amines
Data Sheets

Fill in the data tables with your observations and results (color, solubility, etc.).

Solubility of Alcohols, pH test

Alcohol	Cold Water	Hot Water	pH Test
Ethanol			
t-Butyl Alcohol			XXXXXXXXX
cyclohexanol			XXXXXXXXX
Benzyl Alcohol			XXXXXXXXX

Solubility of Phenols, pH Test

Phenol	Cold Water	Hot Water	NaOH	pH Test
Phenol			XXXXXXXX	
p-Chlorophenol			XXXXXXXX	XXXXXXXXX
2-Naphthol				XXXXXXXXX
Salicylic Acid			XXXXXXXXX	XXXXXXXXX

Reactions of Phenols

Phenol	Bromine Water	Ferric Chloride
P-Chlorophenol		XXXXXXXXXXXXXXXXX
Salicylic Acid	XXXXXXXXXXXXXXXXX	

Reactions of Alcohols, The Lucas Test

Alcohol	Alcohol	Results
Primary		
Secondary		
Tertiary		

Reactions of Alcohols, Chromic Acid Oxidation

Alcohol	Alcohol	Results
Primary		
Secondary		
Tertiary		

Write the equations for the reactions of the primary and secondary alcohols with the chromic acid.

Primary:

Secondary:

Blue Litmus Paper Test of Alcohol Oxidation Product Vapor

Alcohol	Alcohol	Results
Primary		
Secondary		

Reactions of Amines
Fill in the data tables with your observations and results (color, solubility, etc.).

	Solubility	Solubility	Color (pH)	Solubility
Amine	**Cold Water**	**Hot Water**	**pH Paper**	**HCl**
Aniline				
Benzylamine				

Experiment 5
Questions

1. Draw the structures for ethanol, t-butyl alcohol, cyclohexanol and benzyl alcohol. What is the relationship between solubility and the structure of these alcohols?

2. Is there a difference between the pH's of the phenol and alcohol solutions. If so, how do you account for the difference?

3. What is the explanation for the differences in reaction rates of the alcohols with the Lucas reagent?

4. Suppose you had four vials, which contained either benzhydrol [diphenylmethanol, $(C_6H_5)_2CHOH$], 4-nitrophenol, 4-nitroaniline or naphthalene (all are solids). What tests would you use to distinguish between and identify the materials in each vial? Hint! Identify all functional groups present in the structures below, then ask what kinds of reactions does each functional group undergo and whether each is acidic, basic or neutral. Another hint, the nitro group is **not** an acid or a base.

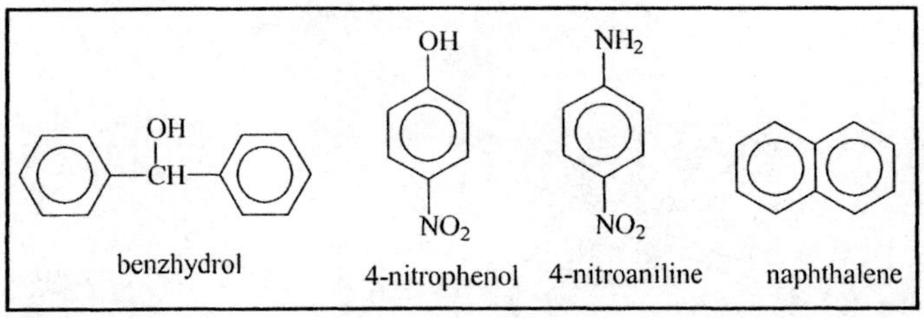

benzhydrol   4-nitrophenol   4-nitroaniline   naphthalene

5. Was there any difference in the basicity of the water solutions of the amines? If so, rank them from most basic to least basic, along with the type of amine ($1°$, $2°$, $3°$, aromatic, heterocyclic aromatic). Ask your instructor if you need help with these definitions.

6. Since pH paper is actually sensitive to the concentration of hydronium ion (H_3O^+) in a water solution, why do the amines turn the pH paper green or blue?

Experiment 6
Properties and Reactions of Aldehydes, Ketones, Carboxylic Acids and Acid Derivatives

I. Aldehydes and Ketones

The aldehydes and ketones belong to a larger family of chemicals which include the carbonyl (C=O) group. The family includes the carboxylic acids and derivatives, such as the esters, amides, acid chlorides and anhydrides. The chemistry of all of these carbonyl compounds are similar, resulting from the polar nature of the C=O bond, as illustrated in **figure 1**. The aldehydes and ketones are very similar in structure and undergo many reactions under similar reaction conditions and therefore are usually considered together when discussing their chemistries. As illustrated in **figure 1**, aldehydes always have at least one hydrogen attached directly to the carbonyl carbon. The second group attached to the carbonyl carbon can be another hydrogen (formaldehyde, methanal) or a carbon group (such as CH_3 for acetaldehyde, ethanal). The ketone functional group always contains 2 carbon groups attached to the carbonyl carbon (acetophenone, methyl phenyl ketone or phenylethanone). We will use 2 chemical reactions which are often used to determine whether or not an unknown contains an aldehyde or ketone (the 2,4-DNP test) and used to distinguish between aldehydes and ketones (the Tollens' or silver mirror test).

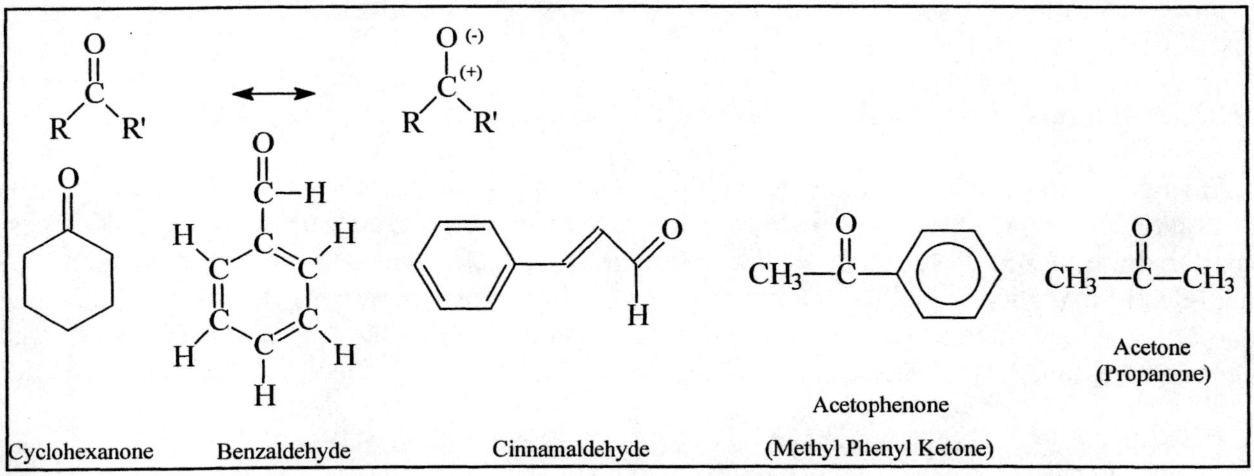

Figure 1. Common Aldehydes and Ketones

Tests for the Presence of an Aldehyde or Ketone Functional Group

A fairly simple, rapid and reliable test to determine whether or not an aldehyde or ketone functional group is present in an unknown material is to treat the unknown with an acidic ethanol solution containing 2,4-dinitrophenylhydrazine (2,4-DNP), as illustrated in figure 2.

97

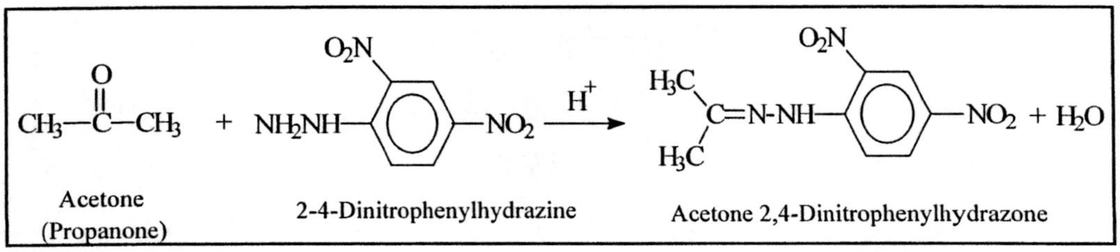

Figure 2. 2,4-DNP Derivatives of Aldehydes and Ketones

The 2,4-DNP reagent is prepared by dissolving the bright orange 2,4-DNP crystals in concentrated H_2SO_4 and diluting the solution with a water/ethanol mixture. The aldehyde or ketone reacts with the -NH_2 group of the 2,4-DNP, resulting in replacement of the oxygen of the carbonyl with the nitrogen of the DNP. The final product is called a 2,4-dinitrophenyl hydra*zone*. The hydrazone ranges in color from yellow to red-tending toward yellow if the aldehyde or ketone has alkane carbon attached to the carbonyl carbon (acetaldehyde or acetone) and tending toward red if a double bond or aromatic ring is attached (benzaldehyde). The hydrazone produced in the test can also serve to aid in the identification of the unknown molecule. Aldehydes and ketones are often liquids. The corresponding 2,4-dinitophenylhydrazones, on the other hand, are always solids. When the solid derivatives are purified by recrystallization, they have sharp melting points which can be compared to tabulated book values for aldehyde or ketone 2,4-DNPs, leading to the identification of the unknown. One word of warning. Occasionally, the 2,4-DNP crystals will precipitate out from solution, leading to a false positive test for an aldehyde or ketone. However, the hydrazone product is usually formed in large amounts and normally easily distinguished from the hydrazine. A melting point will easily determine whether or not the material is a hydrazone.

Tests to Distinguish Between Aldehydes and Ketones

Aldehydes can be oxidized under very mild conditions. In fact, reaction with oxygen in the air will lead to fairly rapid oxidation of many aldehydes to the corresponding carboxylic acid. Ketones usually require vigorous oxidizing conditions, such as hot potassium permanganate solution. A reliable, mild oxidation technique traditionally used to distinguish between aldehydes and ketones is the Tollens' test, often called the silver mirror test. In performing the test, it is necessary to use clean test tubes, since even trace amounts of aldehyde contaminants will lead to a false positive test.

The Tollens' reagent is prepared by first reacting silver nitrate with sodium hydroxide to produce brown silver oxide precipitate, then reacting the silver oxide with ammonium hydroxide solution to yield the silver ammonium hydroxide complex, as illustrated in **figure 3**. The aldehyde then is oxidized by the reagent to an ammonium salt of the corresponding carboxylic acid, while the silver complex is reduced to silver metal. If the aldehyde is present in small amounts (because the solution is dilute or the aldehyde isn't completely soluble in the water) the silver metal forms on the walls of the glass test tube, producing a mirror. If the aldehyde concentrations are high or the test tube is shaken, the silver often precipitates rapidly as a grey precipitate.

This technique has been used in the past to produce silver mirrors on plate glass. Most mirrors today are produced using aluminum instead of silver.

$$2\ AgNO_3\ +\ 2\ NaOH \longrightarrow Ag_2O\ +\ H_2O\ +\ 2\ NaNO_2$$

$$Ag_2O\ +\ 4\ NH_4OH \longrightarrow 2\ Ag(NH_3)_2OH\ +\ 3\ H_2O$$

$$\underset{\overset{\|}{R-C-H}}{\overset{O}{}}\ +\ 2\ Ag(NH_3)_2OH \longrightarrow \underset{\overset{\|}{R-C-O^{(-)}}}{\overset{O}{}}\ NH_4^{(+)}\ +\ 2\ Ag^o\ +\ NH_4OH$$

Figure 3. Silver Mirror (Tollens') Test for Aldehydes

II. Carboxylic Acids

The carboxylic acid also contains the carbonyl group (C=O), but with a hydroxy group (OH) attached to the carbonyl carbon. The hydroxy group, however, is **not** an alcohol. While the carboxylic acids undergo reactions similar to those of the aldehydes, ketones and alcohols, as described below, the most important property of this functional group is the acidic nature of the hydroxy hydrogen. The structures for several common carboxylic acids are shown in **figure 4**. The carboxylic acid functional group can be written as CO-OH or -COOH or -CO₂H.

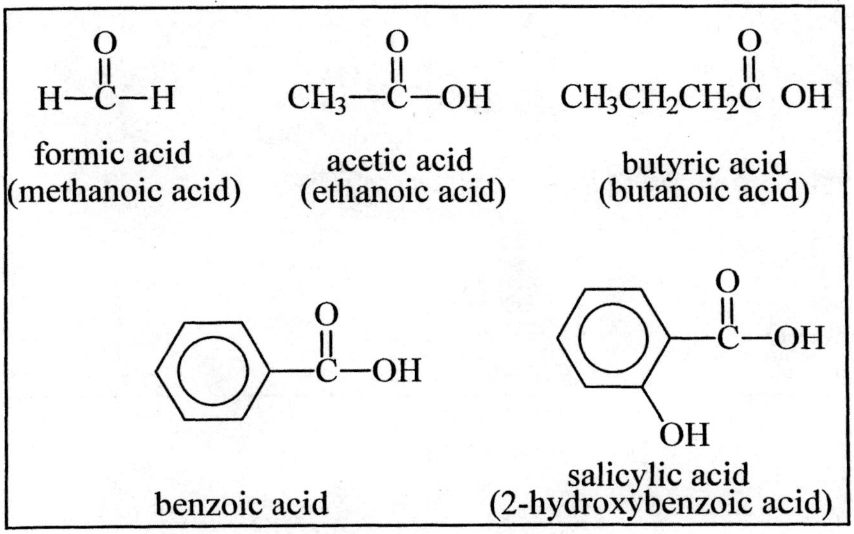

formic acid
(methanoic acid)

acetic acid
(ethanoic acid)

butyric acid
(butanoic acid)

benzoic acid

salicylic acid
(2-hydroxybenzoic acid)

Figure 4. Common Carboxylic Acids

Reactions of Carboxylic Acids

The carboxylic acids undergo two important reactions (1) reaction as an acid and (2) replacement of the OH group to form a carboxylic acid derivative as illustrated in **figure 5**.

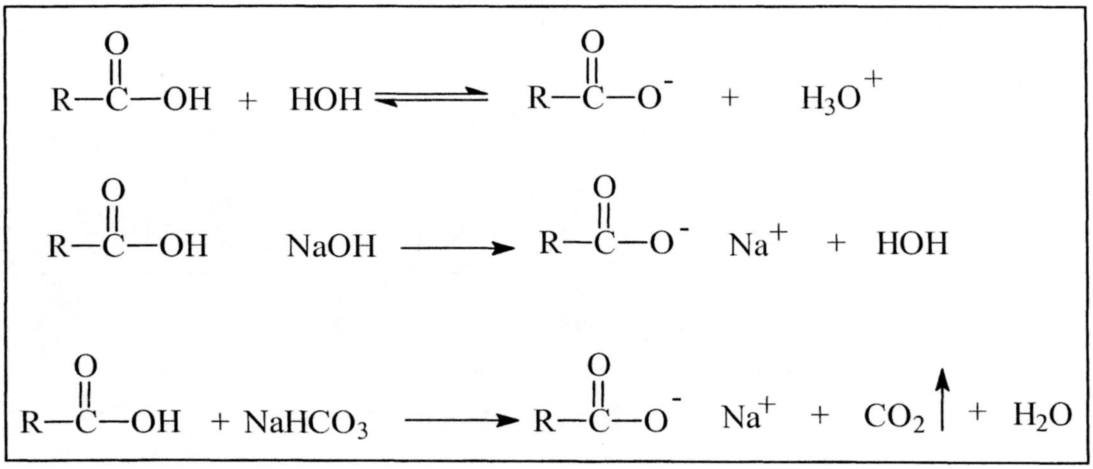

$$R-\overset{\overset{\displaystyle O}{\|}}{C}-OH \;+\; HOH \;\rightleftharpoons\; R-\overset{\overset{\displaystyle O}{\|}}{C}-O^- \;+\; H_3O^+$$

$$R-\overset{\overset{\displaystyle O}{\|}}{C}-OH \quad NaOH \;\longrightarrow\; R-\overset{\overset{\displaystyle O}{\|}}{C}-O^- \; Na^+ \;+\; HOH$$

$$R-\overset{\overset{\displaystyle O}{\|}}{C}-OH \;+\; NaHCO_3 \;\longrightarrow\; R-\overset{\overset{\displaystyle O}{\|}}{C}-O^- \; Na^+ \;+\; CO_2\uparrow \;+\; H_2O$$

Reaction as an Acid

$$\underset{\text{acid}}{R-\overset{\overset{\displaystyle O}{\|}}{C}-OH} \;+\; \underset{\text{alcohol}}{R-OH'} \;\rightleftharpoons\; \underset{\text{ester}}{R-\overset{\overset{\displaystyle O}{\|}}{C}-OR'} \;+\; HOH$$

$$\underset{\text{acid}}{R-\overset{\overset{\displaystyle O}{\|}}{C}-OH} \;+\; \underset{\text{amine}}{R'NH_2} \;\rightleftharpoons\; \underset{\text{amide}}{R-\overset{\overset{\displaystyle O}{\|}}{C}-NH_2} \;+\; HOH$$

$$\underset{\text{acid}}{R-\overset{\overset{\displaystyle O}{\|}}{C}-OH} \;+\; \underset{\text{thionyl chloride}}{SOCl_2} \;\longrightarrow\; \underset{\text{acid chloride}}{R-\overset{\overset{\displaystyle O}{\|}}{C}-Cl} \;+\; HCl\uparrow \;+\; SO_2\uparrow$$

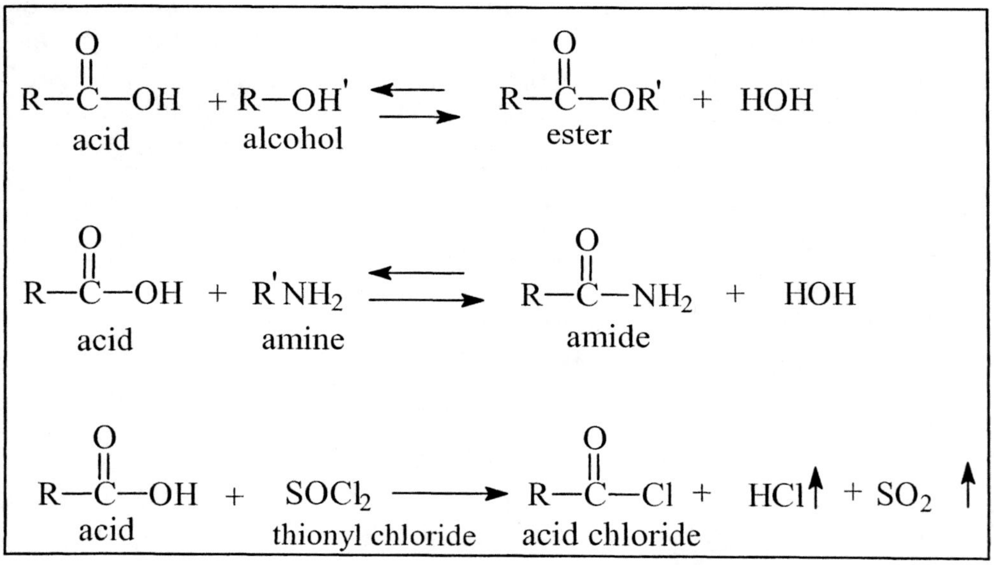

Formation of Acid Derivatives

Figure 5. Reactions of Carboxylic Acids

Since the reaction of a carboxylic acid with base results in the formation of the salt of the base and with water to form H_3O^+, carboxylic acids form an acidic solution (pH=2-5) in water and dissolve in basic solution. In addition, carboxylic acids are strong enough acids to react with the weak base

100

$NaHCO_3$ (baking soda), which, when protonated, decomposes to water and CO_2 gas. Thus, general tests for a carboxyulic acid are: (1) if soluble in water, they form an acidic solution, generally in the range of about pH=3 (2) If water insoluble, it will dissolve in NaOH solution and (3) it will react with $NaHCO_3$ to form a water soluble salt and CO_2 bubbles will be formed.

While carboxylic acids can be directly converted to an ester derivative by reacting with an alcohol or phenol (usually in the presence of H^+ catalyst) or to an amide by reacting with an amine, more commonly they are first converted to an acid chloride (with thionyl chloride, a high yeild reaction), which then is converted to the corresponding ester or amide (also high yield reactions), as illustrated in **figure 6**.

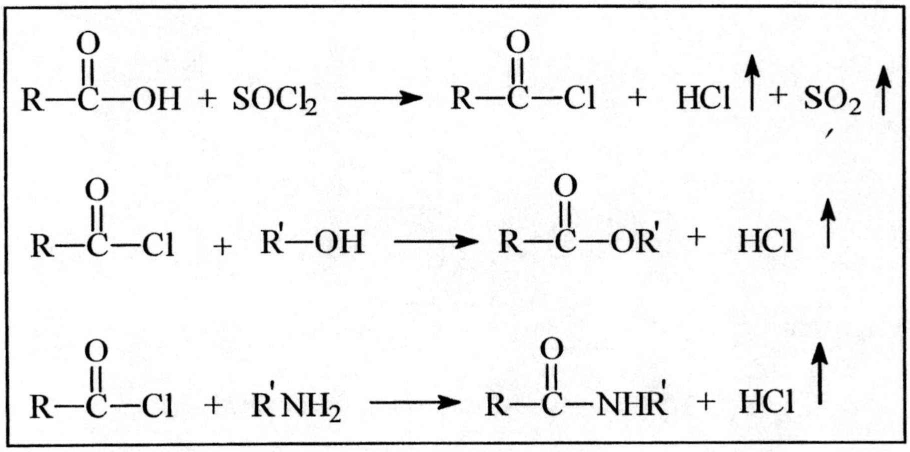

Figure 6. Formation of Esters and Amides from Acid Chlorides

Esters are frequently used in perfumes and as flavorings in foods. We will use benzoyl chloride later in this experiment, to prepare an amide and an ester. In Experiment 7 we will prepare a polyamide (nylon 6,10) using this method.

Finally, it should be mentioned that all carboxylic acid derivatives can be converted back to the carboxylic acid by treating them with water (hydrolysis), in the presence of acid or base, as illustrated in **figure 7**.

101

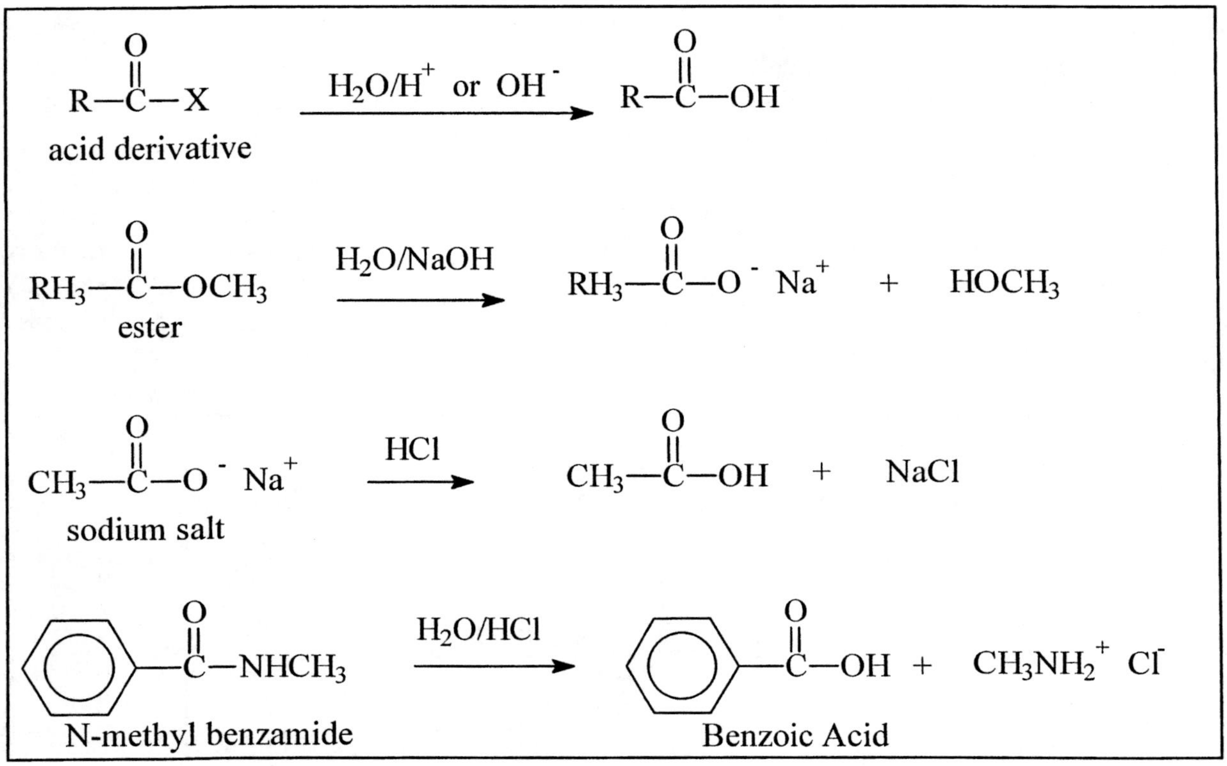

Figure 7. Hydrolysis of Carboxylic Acid Derivatives

Hazardous Materials

The 2,4-DNP reagent is prepared with sulfuric acid. Take appropriate precautions to avoid contact with your skin. Many aldehydes and ketones have pleasant smells when in very dilute form. Many are commonly used for flavorants in foods and drinks (almonds, vanilla, cinnamon). However, they often have very disagreeable odors in concentrated form. In addition, formaldehyde is a suspect carcinogen. Be careful not to spill the chemicals and do not dispose down the drain. Always place in the organics waste container, along with ethanol washes. Silver nitrate, when absorbed on skin or clothing will turn dark brown after exposure to light. This color is permanent. Your clothing and skin will be permanently stained.

Amines, while usually not harmful to the skin, often have bad odors (some smell like rotten fish-a good classification method) and care must be taken not to spill the material or improperly dispose of it in the sink. Always dispose the chemicals in the organics waste container in the hood, rinsing glassware with ethanol. You will also be working with concentrated HCl, which is harmful to the skin, and its vapors harmful to the eyes and lungs. Sodium hydroxide is caustic and harmful to the skin. Benzoyl chloride reacts with water in the skin to form HCl. Always wash your skin or clothing thoroughly if exposed to these chemicals.
MTBE is very flammable. No flames!

Disposal of Hazardous Waste

Follow the general method for disposal of hazardous waste, pouring reaction tube contents into the non-chlorinated organics or chlorinated organics waste bottles, as indicated for each test. Rinse each tube with an additional 2 to 3 portions of ½-1 mL of acetone.

Procedure

A. Test for Aldehydes and Ketones, Reaction with 2,4-Dinitrophenylhydrazine (2,4-DNP)

1) Add 1 mL of 2,4-DNP solution to each of 4 labeled test tubes. Add 2 drops of **one** of the following chemicals, separately, to each test tube: cinnamaldehyde, acetone, cyclohexanone and acetophenone. Mix the contents with the finger flick method. Within a few seconds a yellow to red-orange colored precipitate will form.

2) This material could, if time permitted, be filtered and recrystallized. The melting point of the resulting pure crystals would be indicative of the starting carbonyl compound. Unfortunately, there will not be enough time to do this.

3) Add 2 mL of acetone to each DNP Tube and pour the material in the *organics waste container*, rinsing with two or three 1/2 mL portions of acetone into the *acetone waste container*.

B. Test to Distinguish between Aldehydes and Ketones, The Tollens' Test-Formation of a Silver Mirror

Caution, DO NOT use a Reaction Tube for this experiment. Use the small Test Tube provided for this experiment.

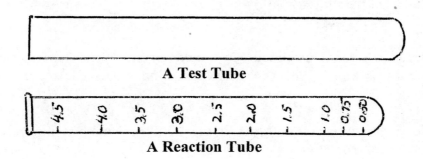

A Test Tube

A Reaction Tube

1) This test is very sensitive. Clean 4 small **TEST TUBES (do not use reaction tubes)** very carefully, with soap and water, if necessary, followed by D.I. water to remove any contaminents in the soapy water. Label each test tube and add 1 mL of a 5% solution of silver nitrate

silver nitrate to each. To each test tube add 1 drop of 10% sodium hydroxide. You should have a brown precipitate in each tube.

2) Begin adding dropwise a 2% ammonia solution to one of the test tubes. You can add 3-4 drops each time at first-thoroughly mixing the test tube using the finger flick technique. When the precipitate begins to dissolve, add ammonia one drop at a time between thorough mixing until the precipitate just dissolves. There may be a few **flecks** of solid remaining in the test tube. This is probably silver metal. Add another drop of ammonia and mix. If the flecks do not dissolve, you have added enough ammonia. Do not add an excess of ammonia or the reaction will not work.

3) Add ammonia to the remaining test tubes in the same fashion. You now know approximately how much ammonia to add, so these should proceed faster.

4) To the four solutions now add separately and carefully **1 drop** of formaldehyde, cinnamaldehyde, benzaldehyde and acetone. **Do not shake the test tubes.**

5) Carefully, to avoid shaking, place the test tubes in a warm (not hot) water bath (40-50°C) for 5 minutes. If you shake the tubes, the reaction will occur so quickly that a grey precipitate of silver metal will form in the test tube. If you are gentle, the reaction will occur slowly and the silver will plate out on the walls of the test tubes and produce a beautiful silver mirror.

When finished, pour the solution into the special waste bottle labeled *silver mirror tests* and leave the unwashed test tubes in the test tube container in the hood.

C. Acid Properties of Carboxylic Acids

1) Place about 10 mg (a small spatula-tip full) of benzoic acid in a test tube. Add 1.0 mL water. Does the acid dissolve? Heat the tube up for 5 minutes in a hot water bath (~70°C) to dissolve the acid. Use a stirring rod to remove a drop of the hot solution and test with a 1/4 inch piece of pH Hydrion paper. What is the pH?

2) Cool the solution to room temperature in a beaker of tap water. The acid should recrystallize. Add dropwise and with mixing 1/2 mL 5% $NaHCO_3$ solution. Note the formation of bubbles and that the acid dissolves.

3) Add, at a drop at a time, concentrated HCl to the tube. Note that, after all the $NaHCO_3$ has reacted, the benzoic acid reprecipitates.

4) Pour the contents of the tube into the *non-halogenated organics waste container* and then rinse the tube two-three times with acetone.

D. Reaction of Amines with Carboxylic Acid Chlorides-Formation of an Amide

1) Place 2 drops of aniline in a clean test tube and add 20 drops of 10% NaOH solution. Mix the contents and add 2 drops of benzoyl chloride to the tube followed by 5 drops of water solution. Use the red cap on the top of the tube and shake the contents. You will observe the formation of the solid amide (benzanilide).

2) The NaOH is added to neutralize the HCl formed during this reaction. The amide derivative could, if time permitted, be filtered and recrystallized. This would yield pure crystalline amide derivative of the amine which would have a unique melting point. This technique is often used to determine the identities of unknown amines. Unfortunately, time does not permit this.

3) Add 1 mL of water to the tube and stir the contents with a stirring rod. Pour the contents into the ***non-halogenated organics waste container*** and rinse once with 1.mL of water. The Cl has been converted to NaCl. Rinse the tube 2 or 3 times with acetone.

E. Reaction of Alcohols with Carboxylic Acid Chlorides-Formation of an Ester

1) Place 5 drops of ethanol and 20 drops of 10% NaOH into a clean test tube. Add 2 drops of benzoyl chloride and then 5 drops of water to the tube. Place a red cap on the tube and shake thoroughly.

2) Let the tube sit for 10 minutes, periodically mixing the contents. Finally, heat the tube (remove the cap) for 5 minutes in a hot water bath. Cool and add ~1/2 mL of the ***blue dyed. water***, to see if an insoluble organic layer is visible (the ethyl benzoate product has a density of 1.05 g/mL). Carefully smell the contents of the tube. The ester is reported to have a "fruity fragrance." What do you think?

3) Pour the contents into the ***organics waste container*** and rinse two-three times with acetone.

Experiment 6
Reactions of Aldehydes, Ketones, Carboxylic Acids and Acid DerivativesData Sheets

Fill in the Data tables with your observations and results (color, solubilities, etc.).

1. 2,4-DNP Test

Aldehyde/Ketone	Results
Cinnamaldehyde	
Acetone	
Cyclohexanone	
Acetophenone	

2. Silver Mirror (Tollens') Test

Aldehyde/Ketone	Results
Formaldehyde	
Cinnamaldehyde	
Benzaldehyde	
Acetone	

3.) Solubility of Benzoic Acid

	Cold Water	Hot Water	pH	NaHCO$_3$	HCl
Benzoic Acid					

4) Formation of an amide

Describe the product. Write the equation for the preparation of benzanilide.

5) Formation of an Ester

What did your ethyl benzoate ester smell like? Write the equation for the preparation of ethyl benzoate.

Experiment 6
Questions

1. If you had an unknown which could be cinnamaldehyde, benzaldehyde or acetone, how would you be able to distinguish between them using chemical tests? Hint 1; Draw the structures for these molecules and identify all their functional groups.

2. How do the above chemicals differ in their appearance and odor?

3. Draw the structures for the 2,4-DNP products obtained in the experiments you performed.

4. Draw the products, if any, of the products obtained in the Tollens' tests you performed.

5. Using reaction tubes, methyl t-Butyl ether, 5% HCl and 5% NaOH, detail how you would separate a mixture of benzoic acid, nitrobenzene and aniline? Hint! Are these chemicals acidic, basic or neutral?

Experiment 7
Polymers
I. Introduction

The word polymer literally means "many (poly) units (mer)." Polymeric compounds include man made products such as the familiar plastics and elastomers we use daily, as well as synthetic oils, paints and adhesives. Natural polymers include natural rubber, cellulose (wood) starch, and protein (yes, we are made up of polymers). There are many specific methods for making polymers, but there are two general chemical classes that most polymers fall into: addition polymers and condensation polymers. Addition polymers are often called polyolefins, since they are usually prepared from olefins (olefin is an old common name for alkene). Thus, ethylene (ethene) is polymerized to yield polyethylene and styrene (phenylethene) is polymerized to yield polystyrene.

Polymers contain from hundreds to thousands of mers linked together, yielding very large and very high molecular weight molecules. Since it would be difficult to display the entire molecule, a shorthand technique is used to draw polymers, as illustrated below. The method consists of drawing the mer(s), with a bond on each side, representing attachment to additional mers on each side. The mer is then enclosed in parentheses and a subscript n placed on the right. This represents a polymer containing "n" mers connected together.

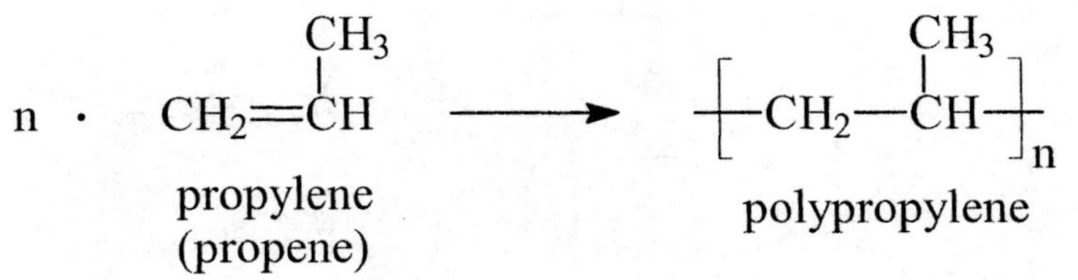

$$n \cdot \quad CH_2\!\!=\!\!\overset{\overset{\displaystyle CH_3}{|}}{CH} \quad \longrightarrow \quad \left[\!\!-CH_2\!\!-\!\!\overset{\overset{\displaystyle CH_3}{|}}{CH}\!\!-\!\!\right]_n$$

propylene (propene)   →   polypropylene

Polyolefins

Polyolefins are usually named by placing the prefix "poly" in front of the common name for the alkene mer. Thus, the polymer of propylene (propene) is named polypropylene. If the mer name is composed of two parts, the mer is placed in parentheses. Thus the polymer of vinyl chloride (chloroethene) is poly(vinyl chloride) and the polymer of methyl methacrylate (methyl 2-methylpropenoic acid) is poly(methyl methacrylate) These names are often shorted to "alphabet soup" names by using the beginning letters for the mer, preceded by a P for poly. Thus polyethylene becomes PE, polytetrafluroethylene becomes PTFE (trade name Teflon), and poly(methyl methacrylate) becomes PMMA (trade name Plexiglass). Other common polyolefin alphabet names are PP (polypropylene), PS (polystyrene), PVA [poly(vinyl acetate or poly(vinyl alcohol)] and PAN (polyacrylonitrile). Despite being named polyethylene, this and other polyolefins do not contain a double bond. The double bond π electrons are used to form the 2 single bonds connecting the mers together.

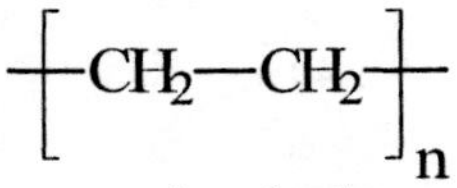

polyethylene 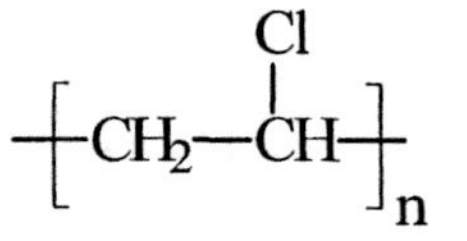   PVC 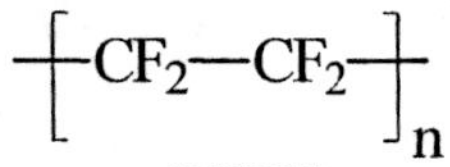   PTFE 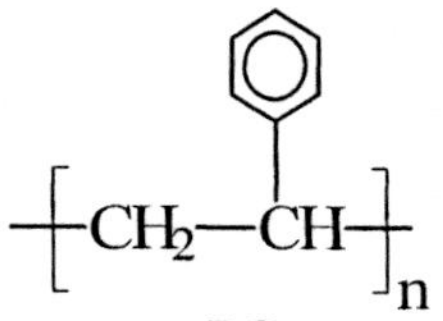   PS

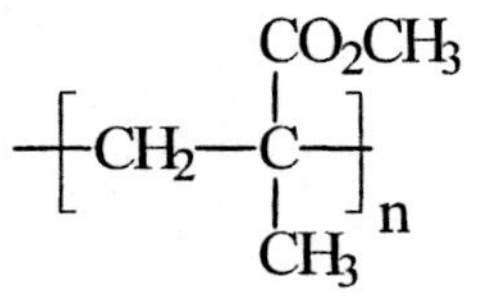

PMMA 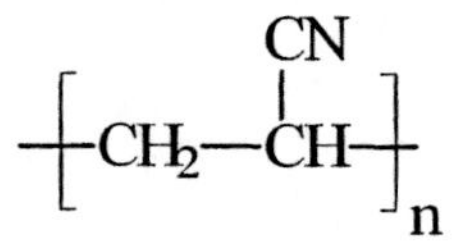   PAN 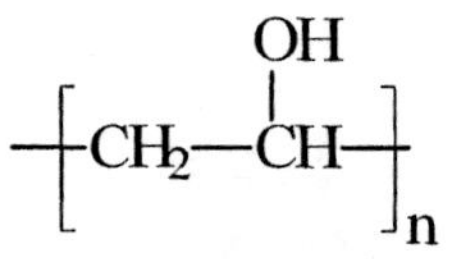   PVA

Elastomers

Polymers can also be prepared from butadiene derivatives. This type of polymer has double bonds in the middle of what was the mer, and are often used as rubbers or elastomers. Natural rubber is a polymer of isoprene (2-methyl-1,3-butadiene). Neoprene rubber is prepared by polymerizing 2-chloro-1,3-butadiene. Two of the 4 electrons in the 2 π bonds are used to connect the butadiene mers together. As a result, the butadiene polymers have 1 double bond in each mer, located between the 2 middle carbons of the original mer, as indicated below.

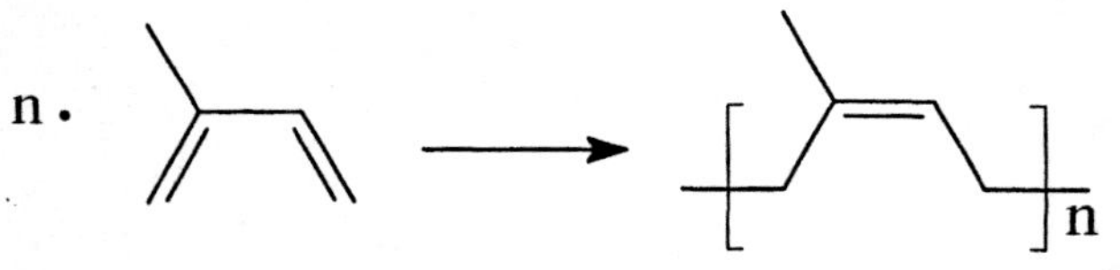

isoprene   cis-polyisoprene   chloroprene   polychloroprene
2-methyl-1,3-butadiene   (latex rubber)   2-chloro-1,3-butadiene   (neoprene rubber)

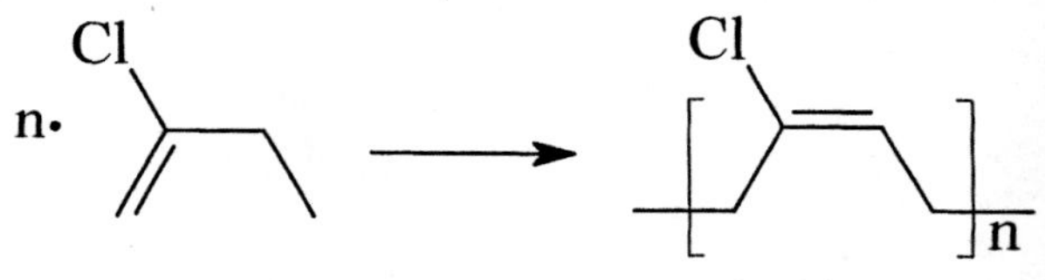

Co-Polymers and Tacticity

Polymers are often prepared by polymerizing together more than one mer. This type of polymer is called a co-polymer. Co-polymers can be prepared with various percentages of each mer or with the mers alternating, random or in block arrangement. Finally, the substituent(s) on the mer (for instance, the methyl attached to ethylene in propylene) can be arranged so that they are constantly on one side of the polymer chain, alternate or are randomly arranged. These arrangements are said to be isotactic, syndiotactic or atactic.

112

X = mer 1

$\{-XXXXYYYXXXYYY-\}$ $\{-XYXYXY-\}$ $\{-XYXXYXYY-\}$

Y = mer 2   **Block Co-Polymer**   **Alternating Copolymer**   **Random Copolymer**

Isotactic PP   Syndiotactic PP   Atactic PP

Each of these arrangements yields a polymer with different chemical, physical and mechanical properties, with differing applications for industrial and consumer use.

Condensation Polymers

The second major category of polymers are the condensation polymers. These polymers are usually prepared by reacting a carboxylic acid or acid derivitive with an amine, alcohol or phenol to form the condensation product, an amide or ester. In order for the product to be a polymer, a di-carboxylic acid or derivative is reacted with a di-amine, di-alcohol or di-phenol to yield a polyamide or polyester. PET is prepared by **P**olymerizing **E**thylene glycol (1,2-ethanediol) with **T**erephthalic acid (1,4-dicarboxybenzene) as indicated below. PET (also sometimes called PETE) gets its name from the common names of each of the mers used to prepare it, preceded by the prefix poly. Most Condensation polymers are given common names or trade names, which have become so familiar that they have become common names. Thus names like nylon 6,6 or Kevlar are commonly used for two important polyamides. We will make nylon 6,10 in this experiment.

ethylene glycol   terephthalic acid   poly[ethylene terephthalate]
(PET)

Thermoplastics and Thermosets

Polymers can also be categorized another way: as either thermoplastics or thermosets. Thermoplastics are polymers which can, at least theoretically, be melted and cast into a new shape. The bonding between thermoplastic polymer molecules is due to weak interactions such as hydrogen bonding and van der Waals forces. Thus, these bonds are not permanent and the polymer will melt when heated. Thermosets, on the other hand, cannot be melted and reshaped after the polymerization process has been completed. The crosslinking of polymer chains for these molecules is normally covalent and thus permanent. In order to break these bonds, the molecule must be heated to the point of decomposition. Thus, these molecules will not melt, but are usable to fairly high temperatures. There are also some crosslinked polymers, such as the slime which we will prepare today, which seem to fit in between these categories where the "crosslinking" is with bonds such as ionic, hydrogen or dipole bonds. These are actually thermoplastics, since the crosslinking is not permanent, but they have some interesting characteristics. A few examples will illustrate the properties of various types of polymers. Polystyrene is a brittle thermoplastic and can be melted over and over and cast again and again into new shapes. Bakelite is a thermoset and once it is cured, it cannot be melted. It decomposes when heated to a high temperature. "Silly Putty" on the other hand, is a thermoplastic silicone which will not hold its shape, but it is very elastomeric.

Plasticizers

Often, chemicals are added to hard polymers to make them soft, flexible and elastomeric. An example is the vinyl plastic used for car seats, shower curtains, etc. Vinyl plastic is actually PVC with plasticizers (usually chemicals called phthalate esters) added. Water can also act as a plasticizer for nylons. We will see this effect for the nylon 6,10 that we prepare. Immediately after its synthesis, it is flexible. After it "drys out", however, it becomes inflexible and brittle.

II. Crosslinking Poly(vinyl alcohol), Preparation of "Slime"

In this experiment, you will "crosslink" poly(vinyl alcohol) with a crosslinking agent, sodium borate (or borax as the decahydrate), which does not form permanent bonds between the molecules, as in Bakelite. In addition, the polymer and the borax also hydrogen bond to water molecules. The resulting material is a soft, gelatenous material, commercially sold as "Slime." The starting polymer, poly(vinyl alcohol) is an interesting polymer in its own right, since it is soluble in water, unlike most other synthetic polymers. This is due to the presence of the alcohol (OH) group which hydrogen bonds to water. The hydrogen bonding is so extensive that the polymer is solubilized in the water.

The sodium borate used in the experiment has the formula $Na_2B_4O_7$. When it is crystallized with water, it forms a decahydrate (known in this form as borax) and some of the water is incorporated into the molecular structure to form a large cyclic molecule with OH groups attached to the borons. The O and H atoms hydrogen bond to the O and H of the alcohols in poly(vinyl alcohol), holding the polymer molecule together, forming an insoluble gel of polymer, water and borax.

III. Synthesis of Nylon 6,10

Introduction

Linear polymers are composed of long chainlike molecules. The chains have been built-up from individual small molecules called monomers. The monomers can be chemically linked to form polymers by one of two general mechanisms; addition polymerization or condensation polymerization.

Addition polymers are prepared by the polymerization of alkenes or dienes. Condensation polymers involve linking monomers which contain two or more reactive functional groups, often forming a small molecule such as water or methanol as a by-product. Today we will prepare a condensation polymer, nylon 6,10 (pronounced "nylon" six-ten). Nylon is a thermoplastic polymer, since it can be melted and molded into a new shape. Bakelite, on the other had, is a thermoset polymer since it cannot be melted and molded. This difference in properties occurs because nylon molecules are separate, individual chains, held together only by the relatively weak force of hydrogen bonding. Bakelite, however, is extensively crosslinked between chains by covalent bonds, resulting in a rigid three dimensional network. The reactions for preparing these polymers are as follows:

1,6-hexanediamine   Sebacyl chloride   Nylon 6, 10
                          (Decanedioyl chloride)

phenol   formaldehyde   + H$_2$O

Bakelite

IV. Identification of an Unknown Polymer

Because the properties for many polymers are similar, it is usually necessary to use more than one method to identify a polymer. We will use a combination of differences in density, solubilities, infrared spectra and thermal properties to identify a polymer unknown in a list of 6 possible polymers.

A. Physical and Chemical Properties of Polymers

Polymers are composed of polymer chains containing various numbers of linked mers. Thus, unlike small organic molecules like ethanol or naphthalene, the physical properties of polymers vary with the average of the molecular weights of the polymer chains, the tacticity of the polymer, crosslinking, and many other factors. For instance,   Low Density PolyEthylene (**LDPE**), is a polymer which contains many chain branches and, the chains cannot approach one another and "crystallize." Thus,

LDPE chains are relatively far apart and have a density of 0.92-0.94 g/cc. **High Density PolyEthylene (HDPE)**, on the other hand, has linear chains that approach one another closely and "crystallize" with a density of 0.95-0.97 g/cc. This difference in density is used to separate these polymers in recycling plants. The different physical properties of polymers also gives rise to different uses for the polymers, since they also have different mechanical properties, different resistance to chemical attack, and some are crystal clear and others opaque. We will use physical and chemical differences, including density, chemical resistance and combustion differences, to identify the 6 polymers, listed below.

Recycling #	Chemical	Density (g/cc)
	Water	1.00
1	Poly(Ethylene Terephthalate) (PET)	1.38-1.39
2	High Density Polyethylene (HDPE)	0.95-0.97
3	Poly(Vinyl Chloride) (PVC)	1.16-1.35
4	Low Density Polyethylene (LDPE)	0.92-0.94
5	Polypropylene (PP)	0.90-0.91
6	Polystyrene (PS)	1.05-1.07

The **recycling numbers** are used on plastic parts to identify them for recycling. The numbers were assigned to the six most widely recycled polymers with PET (pop bottles) being most commonly recycled.

Chemical resistance is also useful for identifying polymers. We will use the partial solubility of polystyrene in acetone to distinguish PS from PET, which is quite resistant to chemicals.

Finally, combustion or burning characteristics can also be a useful tool to identify polymers which contain chlorine. As we saw in experiment 2, PVC burns with a green flame in the presence of copper metal.

B. Infrared Spectroscopy

One physical characteristic of chemicals is their absorption of infrared (IR) light at specific wavelengths, giving rise to a so-called infrared absorption spectrum that is unique for each different chemical. The Fourier Transform Infrared Spectrometer is used to obtain an IR spectrum of chemicals and identify them. This technique is more fully explained in Experiment 8. Each polymer also has its unique IR spectrum which can be used to identify it. Since the IR spectrum results from the unique absorption of IR light by organic functional groups, when polymers have very similar structures, such as LDPE and HDPE, they may have very similar IR spectra, so additional techniques may also be needed to identify the polymer. We will not try to interpret individual absorptions in the IR spectrum in this experiment. We will instead identify a polymer by comparing the total pattern of the spectrum for an unknown polymer to the spectrum patterns for known polymers.

C. Thermal Analysis: Differential Scanning Calorimetry (DSC)

DSC is a thermal method used to identify polymers and to determine two very important polymer processing and polymer use properties: glass transition temperature (T_g) and melt temperature (T_m). Polymers don't behave like small molecules, which normally melt over a small temperature range. Polymers are usually a complex mix of polymer chains of varied lengths, and are thus really a mixture. If the polymer chains can bond to one another (by van der Waals, hydrogen bonding, etc.) extensively, the polymer is said to be crystalline. HDPE and PP are examples of crystalline polymers. If the polymer chains cannot bond together in a regular manner, but instead are randomly arranged in the solid, the polymer is said to be amorphous. PS and PVC are amorphous polymers. Often polymers have partial crystallinity and partial amorphous nature. PET and LDPE are such polymers.

Crystalline polymers, when heated, tend to melt, although over a more broad range of temperatures than small molecules. The average point, approximately, in the melting point temperature range is referred to as the *melt temperature* or T_m. This temperature is obviously very important from a processing standpoint because is determines the temperature required to melt and form plastics.

Amorphous polymers normally also soften to a viscous liquid which can be forced into a mold to shape the plastic. The temperature range for these polymers to change to a liquid is, however, usually much broader than for crystalline polymers. Amorphous polymers also undergo another unique transition, however, called the glass transition. This usually occurs over a narrow range of temperatures referred to as the **glass transition temperature** or T_g. The T_g is essentially the temperature above which the polymer converts from a hard, inflexible material, to a rubbery or leathery material. The T_g for PS is reported to range from 74-105°C. The T_g, like the T_m, often depends on whether or not additives are present in the polymer. The T_g is important for the mechanical uses of a polymer. As an example, PS certainly could not be used under the hood of an automobile for a structural application. On the other hand, it is an excellent polymer for common commercial applications, such as jewelry cases or model airplanes.

The T_m and T_g are thermal transitions, like the boiling point and melting point for small molecules. The T_m and T_g for polymers are determined with an instrument called a **Differential Scanning Calorimeter (DSC)**. The sample (~5-10 mg encased in a small aluminum capsule) is placed on the instrument furnace and heated through the T_m or T_g transition. At the transition, additional heat is necessary to allow the furnace to maintain a constant temperature rise. The additional electric current required to raise the temperature of the furnace is translated as the DSC thermal curve or thermograph.

The T_m and T_g transitions have unique shapes. The T_m appears as a peak (or represented as a trough if an inverted format for plotting is used) while the Tg appears as a step up (or down). The literature T_m and T_g values for the six polymers we will be using in this experiment (HDPE, LDPE, PP, PVC, PS and PET) as well as the T_g for poly(methyl methacrylate) (PMMA), which will be used to demonstrate the use of the DSC. Example thermographs for PMMA (T_g) and Nylon 610 (T_m) are provided at the end of the experiment.

Polymer	T_m (°C)	T_g (°C)
HDPE	130-137	---------
LDPE	98-115	-25
PP	160-175	-20
PVC	---------	75-105
PS	---------	74-105
PET	245-265	73-80
PMMA	---------	85-105
Nylon 6,10	~220	---------

In summary, DSC is a very important method for determining the processing and use temperatures for polymers, as well as a method for quality control of and determining the identities of polymers and polymer blends. *However, since the T_m and T_g values for these polymers are often similar, additional methods, such as IR, must be used in conjunction with DSC for the identification of polymers*

Procedure

Hazardous Materials

No hazardous materials are used in the slime experiment. However, the food coloring will stain your hands and clothes. Use only 1 drop of food coloring. Wear gloves when handling the polymer. The nylon experiment involves the use of an acid chloride, which can release HCl if it contacts your skin. Wear gloves or use care when dispensing this material. The hexane used in the nylon experiment will dissolve the wax sealant used on the floor and bench top. NEVER PLACE THE NYLON ROPE DIRECTLY ON THE FLOOR OR BENCH TOP WHILE WET, IT WILL STICK WHEN THE SOLVENT DRIES. USE PAPER TOWELING AS INSTRUCTED. No hazardous materials, other than hexane and acetone are used in the polymer identification experiments. A flame is used for the halogen test, however. Handle the copper wire only by the cork holder.

A. Crosslinking Poly(vinyl alcohol), Preparation of "Slime"

Each student will perform this experiment.

1) Place 20 mL of Poly (vinyl alcohol) (PVA) solution in a 100 mL beaker or provided plastic cup and add **ONE DROP** of food coloring. PVA is represented in polymer notation as:

$$\left[-CH_2CHOH- \right]_n$$

2) Add 4 mL of sodium borate solution.

3) Stir the mixture with a stirring rod until a thick gel forms.

118

4) Wearing rubber gloves, so the food coloring does not stain your skin, remove the gel from the beaker and observe its properties. Is it a solid? Does it have liquid properties? Does it have elastic or rubbery properties? Will it "break" if pulled apart rapidly?

5) Place the sample inside one of your gloves or leave it in the plastic cup, capped so it doesn't dry out. You may take it with you, since it has very low toxicity.

*One word of warning -**Don't drop the sample on a rug if it drys out. It becomes hard and the edges will be very sharp and could cut you if you step on it.***

B. Synthesis of Nylon 6,10

1) Place three 5 foot lengths of paper towling on the end of your bench, upon which you will lay the nylon as it is formed. ***DO NOT LAY THE NYLON DIRECTLY ON THE FLOOR OR BENCH.*** The solvent on the nylon will dissolve the wax on the floor or sealant on the bench and the rope will stick.

2) In a dry 50mL beaker, dispense 15mL of a solution containing 3% v/v sebacoyl chloride dissolved in hexane.

3) In a 250 mL beaker, dispense 15 mL of an aqueous solution containing both 1,6-diaminohexane (0.5M) and sodium hydroxide dissolved in water (0.5M).

4) Carefully, and slowly, pour the hexane solution down the side of the 250 mL beaker onto the surface of the water solution. Hexane is less dense than water and will float on top.

5) At the interface of the two layers, the reagents will react and form the polymer. Carefully place the beaker on the bench on the paper toweling and using a pair of tongs or tweezers, reach into the water and grasp the polymer film. Raise the tongs up in a continuous motion. Have your partner grasp the rope with gloved hands and continue to pull the rope up while you place the initial portion of the rope onto the paper towling. Be careful to not let the rope touch the side of the beaker or it will stick & break. If the rope should break accidentally, grasp the film again and continue until no more rope forms. When you can no longer form additional rope, vigorously stir the reagents with a glass rod to form a mass of nylon.

6) Remove the nylon ball from the beaker with a spatula and pat it dry with paper towels. Pour the hexane-water mixture into a waste container labeled NYLON RXN WASTES.

7) Pat the nylon rope dry with paper toweling. Examine the nylon rope and ball for physical properties: Hardness, color, elasticity, etc. If you want, you can measure the length of your rope and compare it to those of your lab-mates.

8) Break off three 1" pieces of your rope and place them in three test tubes. Add 1 mL each of toluene to the first, methanol to the second and concentrated HCl to the third. Place the tubes in a hot (~50° C) water bath for five minutes. Discuss the results with your instructor.

9) Break a 1" piece of the rope off and heat it on a piece of aluminum foil on your hot plate until it softens. Touch the melted nylon with a glass rod and see if it can be deformed. Break off a 2-3" piece of the rope to turn in with your report.

10) After the Nylon sample has dried for at least a day, compare its flexibility and elasticity properties to the nylon sample immediately after its synthesis.

C. Identification of Polymers by Their Chemical/Physical Properties

We will perform the density/solubility/flame tests on six identified polymers. We will simultaneously perform these tests, as needed, on an unknown polymer sample. Each of the known polymer samples has a unique shape/color to help you keep track of its identity during the tests.

Before you begin the tests, record on your data sheet, the size, shape and color for each of the known polymer samples.

Be certain you can distinguish between the pellets before you begin the experiment.

You will also be provided with an unknown polymer pellet, which will be one of the 6 polymers we will be testing. Perform the density/solubility tests on the unknown as well. Stop the tests for the unknown when you feel you have identified it. We will also use FTIR and DSC to identify this unknown polymer. Please note: the unknown pellet will <u>not necessarily</u> have the same size, shape or color as the known polymer pellet.

Follow the flow sheet below, during the identification process.

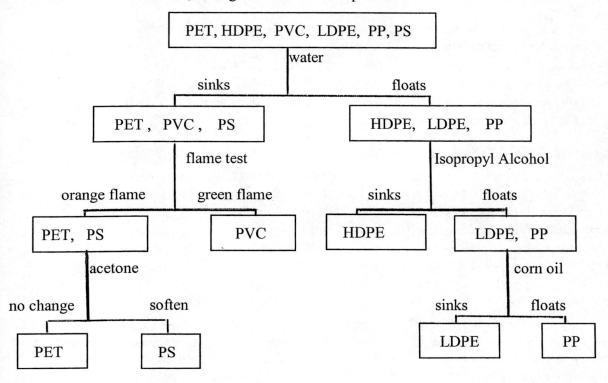

1) Add the 6 polymers to a test tube containing 2 mL of water. Stopper the tube with a red cap and shake the tube vigorously to remove any air bubbles adhering to the polymer pieces. PET, PVC and PS sink, while HDPE, LDPE and PP float in the water. Check the densities for these polymers in the introduction to this experiment. Test the unknown in a second tube, to determine which test to perform on it next.

2) Pour the contents into a 50 mL beaker and remove the polymer pellets with a pair of tweezers. Pour the water in the sink, and dry the beaker and pellets with a paper towel. Add the 3 pieces which had densities less than 1.0 g/cc (floated on the water) to 1 mL of a dilute isopropyl/water solution (the solution has been prepared for you by adding 35 mL of water to 65 mL of 70% isopropyl {rubbing} alcohol). The solution has a density of ~0.95 g/mL so the HDPE will sink while the LDPE and PP will float. Because we are using recycled polymers for this experiment, the densities of individual polymer pellets can vary slightly.

3) If all of the pellets float, add 1 drop of *undiluted* isopropyl alcohol to the test tube and shake. If they all continue to float, continue add 1 drop portions of undiluted isopropyl alcohol (usually 2-3 drops is enough) until the density of the solution is adjusted to the point where one pellet sinks. If all of the pellets originally sank, add one or more drops of water instead of isopropyl alcohol. *Obviously, a delicate touch is needed in adjusting the density of the solution, because the densities of these polymers are very similar. This is why we need additional information, such as an IR spectrum or*

DSC thermograph to accurately identify these polymers.

4) Pour the solvent into a second test tube, leaving the pellets in the first tube, if you need to test you unknown pellet (shake the pellets out of the tube and dry them for the next test). Add the polymer unknown to the liquid to determine its density, if it floated on water in the first test. Pour the solvent into the 50 mL beaker and remove the pellet and pour the isopropyl alcohol into the ***non-halogenated hydrocarbons waste container***. Dry the beaker and pellets with a paper towel. After the pellets have dried for at least 15 minutes, they may be returned to the correct known polymer pellet vials.

5) Add the 2 dried pellets that floated to a ***clean, dry*** reaction tube containing 1 mL of corn oil, which has a density of ~0.91 g/mL. Shake the capped tube vigorously. The LDPE will sink while the PP will float. Pour the corn oil into a second tube if you need to test the unknown, or otherwise pour the contents into the 50 mL beaker and remove the pieces. Pour the corn oil into the ***non-halogenated waste container***. Wipe the beaker with paper toweling, which can be discarded in the waste basket, and then rinse the beaker, pellets and tube ***twice with hexane*** (stopper the tube so you can it) into the ***non-halogenated waste container***. Rinse the tube , pellets and beaker two more times with ***acetone***. Dry the beaker and pellets with paper toweling.

6) The 3 polymer pellets which were heavier than water will now be tested. A copper wire affixed to a cork will be used to flame test the polymers for the presence of chlorine. Heat the end of the wire in a Bunsen burner flame until the wire is red hot. Touch the hot wire tip to one of the pellets to melt some onto the wire. Now place the wire tip into the flame and note the color of the flame. If the polymer does ***not contain chlorine***, it will burn with an ***orange flame***. If the polymer burns with a ***green flame***, it contains ***chlorine***. This is PVC.

7) The last test is a solubility test. Place the 2 remaining polymer pellets into a test tube containing 1 mL of acetone. Perform this test in a second tube on your unknown, if necessary. After 5 minutes, pour the tube contents into the 50 mL beaker. Use a stir rod to check each pellet to see if it has been softened by the acetone. Only the PS polymer is partially soluble in acetone. Remove the pellets and pour the acetone into the ***non-halogenated waste*** container. If necessary, rinse the reaction tube twice with acetone. Dry the beaker and pellets with a paper towel. ***Discard the PS pellet and your unknown pellet, if it is PS,*** into a container labeled ***Solvent Test Pellet*** in the hood. Properly identify all pellets by their color and shape and return to their storage containers.

D. Infrared Spectroscopy and DSC

1) Your instructor will demonstrate how to obtain an IR spectrum and the DSC thermograph using poly(methyl methacrylate) (PMMA, Plexiglass) as an example. You will obtain a spectrum of your polymer unknown in the form of a film taped to a cardboard sample holder. The holder is placed into the IR beam for analysis. Due to time restrictions, you will be provided with a DSC thermograph for the polymer unknown. Identify your unknown using your density/solubility test data and by comparing the IR spectrum and DSC thermograph for your unknown polymer sample to the IR spectra at the end of this experiment and the T_g/T_m data (above) for each polymer.

Experiment 7
Polymers Data Sheet

Attach a 2" piece of your Nylon sample here with tape

Nylon Solubility/Reaction Observations
    Methanol_____

    Toluene_____

    HCl_____

Nylon Physical Properties (Wet)
    Flexibility/Elasticity_____

    Affect of Heat_____

Nylon Physical Properties (Dry)
    Flexibility/Elasticity_____

Slime Properties

    Elasticity_____

    Deformability_____

Size/Shape/Color of the known polymer pellets

    1. PET _____

    2. HDPE _____

    3. PVC _____

    4. LDPE _____

    5. PP _____

    6. PS _____

Draw the solubility/density flow sheet for your unknown polymer

FTIR/DSC Data
Closest polymer match to the FTIR Spectrum for the Unknown_____

T_g _____ or T_m for Unknown_____

Identity of the Unknown Polymer
Unknown Number_____   Unknown Identity_____

Experiment 7
Questions

1. Describe as fully as possible the physical properties of your Nylon rope.

3. What are your observations and conclusions as to the solubility/reactions of Nylon 610?

4. Is your Nylon a thermoset or thermoplastic polymer? Explain.

5. Could this process be used commercially for the production of usable Nylon? Why?

6. Describe the properties of your Slime.

7. What is the identity of your unknown polymer. Explain your reasoning based on the solubility and density tests, the IR and the DSC data.

Infrared Spectra of Polymers

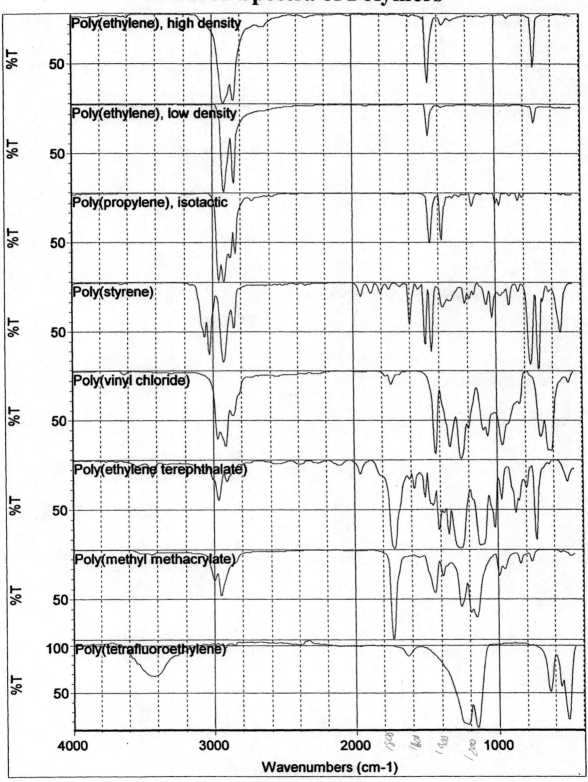

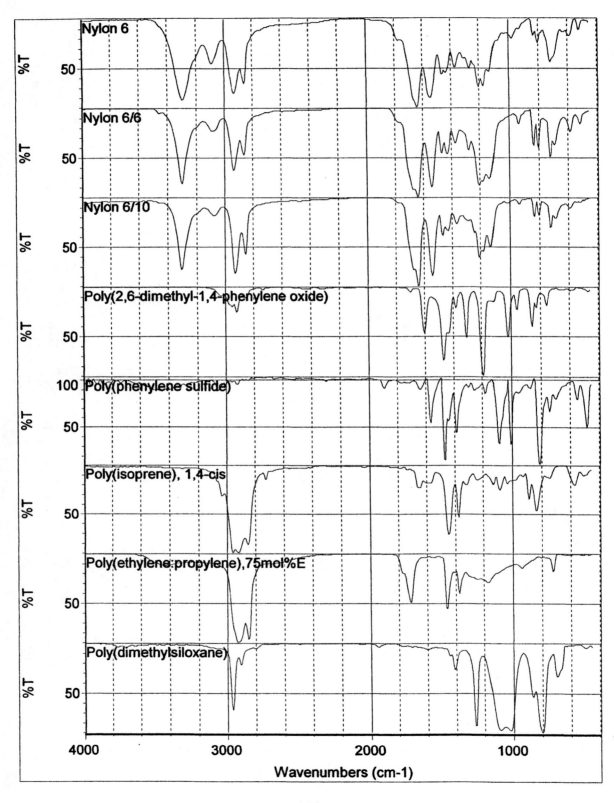

Nylon 6

Nylon 6/6

Nylon 6/10

Poly(2,6-dimethyl-1,4-phenylene oxide)

Poly(phenylene sulfide)

Poly(isoprene), 1,4-cis

Poly(ethylene propylene),75mol%E

Poly(dimethylsiloxane)

Wavenumbers (cm-1)

128

DSC Thermogram for PMMA

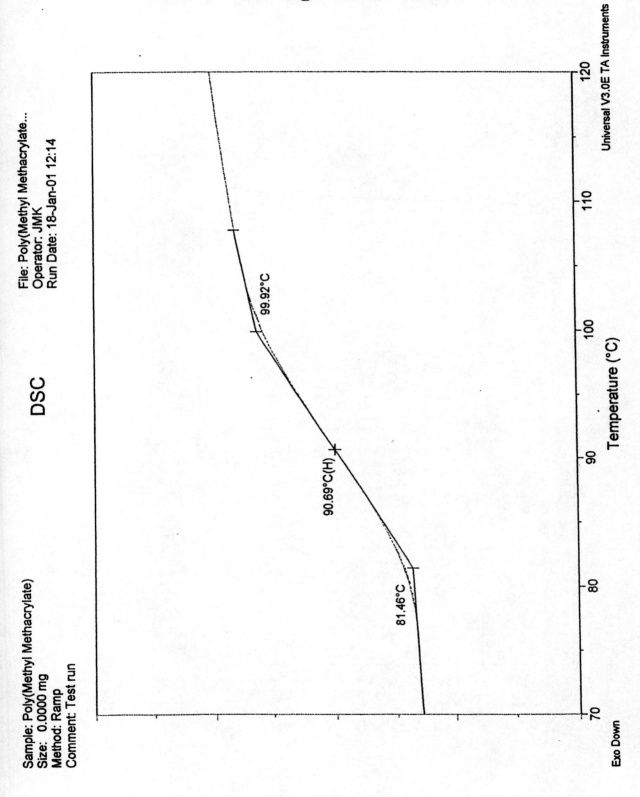

DSC

Sample: Poly(Methyl Methacrylate)
Size: 0.0000 mg
Method: Ramp
Comment: Test run

File: Poly(Methyl Methacrylate...
Operator: JMK
Run Date: 18-Jan-01 12:14

99.92°C

90.69°C(H)

81.46°C

Temperature (°C)

Exo Down

Universal V3.0E TA Instruments

DSC Thermogram for Nylon 6,10

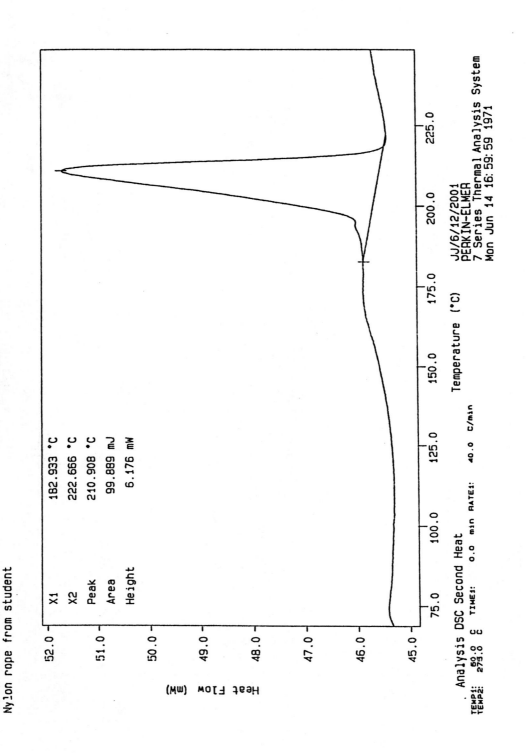

Curve 1: DSC
File info: JJ61201f Sat Jun 12 15:07:03 1971
Sample Weight: 0.000 mg
Nylon rope from student

X1	182.933	°C
X2	222.666	°C
Peak	210.908	°C
Area	99.889	mJ
Height	6.176	mW

Heat Flow (mW)

Temperature (°C)

Analysis DSC Second Heat

TEMP1: 50.0 C TIME1: 0.0 min RATE1: 40.0 C/min
TEMP2: 273.0 C

JJ/6/12/2001
PERKIN-ELMER
7 Series Thermal Analysis System
Mon Jun 14 16:59:59 1971

Experiment 8
Identification of an Unknown

I. Introduction

Most students naturally assume that the purpose of an experiment such as this is to "identify the unknown correctly." While this is true, most of the time the meaning of this phrase is misunderstood. The real purpose of the experiment is to understand how to correctly use the chemical, physical, and spectral data to identify an unknown. In other words, it is the "process" of determining the unknown, and not the exact structure of the unknown that is important. It is of absolutely no use to determine the unknown structure if there is no understanding of how one got it. This experiment has been designed to encourage the development and use of the thought processes used to solve materials characterization problems that will frequently be encountered in the workplace. The major difference between the lab experience and the "real world" is that your unknowns are pure chemicals and not complex mixtures.

II. Take-Home Assignment

Assignment 1: Introduction to Identification of Organic Compounds (IOC)

For this assignment, you will use the IOC program to qualitatively identify organic compounds. Each pair of students will be assigned 1 or more IOC unknowns 8th week. *It is imperative that you and your lab partner work through the following step by step procedure with benzoic acid using the IOC program, before starting to work on your liquid unknown in the lab.* Please read carefully the remainder of this experiment to obtain the necessary background information, however. Turn in the computer program lab notebook for each unknown. *Draw the structure and write the name of the unknown at the end of the lab notebook report.*

Work step by step through the following example. We will examine the logical sequence of tests one would perform to identify benzoic acid.

Benzoic Acid

Knowing the identity of this "unknown," let us walk through the steps by which you will use to identify a compound. Refer to your lab manual and to the on-line help screens at each step for a more complete understanding of the steps and the reactions.

Step 1: Load the IOC Program

From the Windows desktop, select Programs, IOC and then Qual. Org. When the program loads, a Lab Notebook will be displayed, in which all of your test data will be written. This lab notebook will be printed out and included with the unknowns report for Exp 8. To begin an analysis, select New Generic Unknown from the File Menu. A set of test tubes, with unknown numbers appears. Scroll to **test tube 7** and select it, followed with a click on OK.

The program will display physical data for the unknown in the notebook. For benzoic acid, this will be: "White solid, melting point = 118-121 degrees." Note that the melting point is a range rather than a single temperature. This will be the case as well for your solid unknown in lab. (Note: Ignore the statement "You have 5 grams of unknown and 1200 points." You may perform as many tests as necessary to logically determine the structure of your unknown).

The physical data will also include the percentage of Carbon and Hydrogen in the molecule (other unknowns may also include % nitrogen). We will use this data later to calculate the molecular formula and number of double bond equivalents (DBE) for the unknown.

Step 2: Solubility Tests

The solubility tests are the first tests which are normally run on a new unknown in the laboratory, and the same is true for the IOC program. The solubility tests will allow you to classify the unknown as a moderate/strong acid (soluble in 5% $NaHCO_3$ solution, often a carboxylic acid, the IOC program does not contain any sulfonic acid unknowns), as a weak acid (soluble in 5% NaOH solution, usually a phenol), as a base (soluble in 5% HCl solution, an amine) or as a neutral compound (all other common functional groups). In addition, a lower molecular weight compound will be more soluble in water than a higher molecular weight compound, given the presence of same functional groups (ethanol is soluble in water, 1-octanol is only very slightly soluble in water). The presence of double bonds, O, N, or sulfur is implied from the solubility of a neutral compound in sulfuric acid. Three classes of compounds, alkanes, aromatics, and alkyl halides, are classified as "inert" (unreactive), and will give a negative result for all of the solubility tests.

From the Experiments Menu, Select Solubility Tests. The solubility data will be displayed in the lab notebook.

Click on Help and Then Help on Solubility Tests. An appropriate help screen is always available for the test you are working on, by clicking on the last menu item in the help menu. A solubility "tree" or flow chart is displayed for the solubility tests. Carefully examine the solubility tree to see how certain classes can be distinguished from others. Note that the solubility tree is analogous to the "binary" or "decision" trees encountered in computer programming. Beginning with the first test, the solubility of the unknown in water, each test is essentially a "true-false" question. Thus there are tests which you would not perform, based upon which branch of the tree you followed. The help screen, unfortunately, can not be directly printed from the IOC program. To do so requires a page copy (ALT Print Scrn) and paste into another program, such as Word or Word Pad.

Compare the IOC results for Unknown 7 with the Solubility Tree. Beginning at the top of the list, proceed through the solubility tree:

> A. Insoluble in water: The unknown is therefore not a low molecular weight compound. *Had the result been positive, you would have stopped the solubility tests at that point and proceeded to the fusion tests. None of the other tests would be performed, since the compound would be soluble in all aqueous solutions, regardless of pH.*

> B. Proceed to the NaOH test. Soluble in 5% NaOH: The compound is soluble in strong base. The implication here is that the compound is an acid. At this stage we cannot tell whether it is a weak or a strong acid.(Note that a positive test at the NaOH branch means that the HCl and sulfuric acid tests are unnecessary. So we can ignore the results for the latter two tests, even though IOC lists test results for them.)

> C. Proceed to the sodium bicarbonate ($NaHCO_3$) test.
> Soluble in 5% $NaHCO_3$: The compound is soluble in weak base. The implication of this test is that the unknown is a moderate to strong acid, i.e. a carboxylic acid, and some phenols.

So, let us recap what we have learned from just the solubility tests:

1. We know that the unknown is not a low molecular weight compound (i.e. that there are probably more than three or four carbons in the molecule).

2. We have also determined that the compound is a moderate to strong acid, and have thus narrowed the possibilities to a carboxylic acid and some phenols.

Step 3: Sodium Fusion Tests.

The Sodium Fusion Tests allow one to determine whether a compound contains N, S, or halogens. You have not performed these tests in the lab because the first step, the fusion of the sample, takes place at high temperature using sodium metal, and can be dangerous.

Once the sodium fusion step has been completed, three main tests are performed on the fusion mixture to determine the presence or absence of N, S, or halides (Cl, Br, or I). A brief description of each test is given here. After you have performed the initial tests, read carefully the help file for this section to understand the details.

Test 1: For nitrogen. If nitrogen is present, reaction of the fusion mixture with ferrous ammonium sulfate and potassium fluoride in basic solution will produce the cyanide ion (CN⁻) if C-N bonds are present in the molecule. Upon acidification with sulfuric acid, a positive test for N will be a bright blue solution and the formation of a precipitate.

Test 2: For sulfur. A portion of the fusion mixture is acidified and treated with lead acetate. If sulfur is present, insoluble lead sulfide (PbS) will form and a black precipitate will be observed.

Test 3: For halides. Halides form highly insoluble precipitates with silver. The color of the precipitate and its solubility in a concentrated ammonia solution will often allow you to distinguish between chloride, bromide or iodide. Additional "Halide Differentiation Tests" can also be performed if necessary (See the help file for specific information on these tests).

Carboxylic acids will also form precipitates with silver. The presence of a carboxyl group on the molecule may therefore give a "false positive" halide test. A carboxylic acid can be distinguished from a halide by the solubility of the silver carboxylate precipitate in ammonia and concentrated nitric acid. The silver carboxylate will not dissolve in ammonia, but will dissolve in conc. HNO_3.

Perform a sodium fusion test on Unknown 7 by clicking on the Experiments menu, followed by Na Fusion Tests. You will see that a white precipitate is formed when silver nitrate is added to the fusion mixture. Note that the precipitate is not affected by ammonia but dissolves in HNO_3.

Click on the Help menu and then help on the Na fusion tests. The help screen gives a detailed description of all of the chemistry involved in the sodium fusion tests.

Interpret the test results for Unknown 7.

Fusion Test Results for Unknown. #7	Implication of Test Results
Test 1: There is no observable change in the solution (no color change and no precipitation).	No carbon-nitrogen bonds are present in the molecule. Therefore the molecule cannot be an amine nor an amide. (We had already concluded that the compound is not an amine, based upon the solubility tests).
Test 2: No observable change (no precipitation).	No sulfur is present in the molecule. This rules out the possibility of the unknown being a sulfonic acid.

Test 3: A white precipitate is formed.	There is either a halide or a carboxylic acid group in the molecule. If a halide, the white color suggests chloride.
No visible change with ammonia.	The precipitate is not silver chloride (AgCl is soluble in ammonia). Silver iodide is insoluble in ammonia, but it is a yellow precipitate, not white. At this point it appears we have a "false halide test", implying a carboxylic acid.
The precipitate dissolves in nitric acid.	The precipitate is a silver carboxylate, not a silver halide.

Let's review what we know to date. We have a white solid, with melting point of 118-121 degrees Celsius. It is soluble in strong and weak base, implying that it is a carboxylic acid or possibly a phenol. From the fusion tests, we know that the molecule does not contain nitrogen, sulfur, or a halide. The absence of N, S or halogen rules out amines, amides, cyanides, thiols, sulfonic acids or any halogenated compound. A "false positive" halide test indicates that the compound is a carboxylic acid. We have therefore identified a major functional group in the unknown molecule.

Step 4: Molecular Formula.

At this stage, we have enough information to determine a chemical formula for our compound.

For Unknown 7, we are told that the compound is comprised in part of 68.84% C and 4.95% H. Since we already know, from the fusion tests, that there is no N, S, or halide, we can assume that the remaining mass is due to oxygen (Oxygen is always obtained by difference in elemental analysis tests). Therefore the compound is presumed to contain 26.21% oxygen, in addition to the carbon and hydrogen.

For each element, divide the % mass by the atomic weight of that element. This will give the relative number of moles of each element in the compound. For example, $68.84/12.01 = 5.74$ moles of carbon. Now do the same for the hydrogen and oxygen.

For Unknown 1, these calculations give a molecule with the following molar ratios:

$C = 5.74$ $H = 4.90$ $O = 1.64$

Divide each value by the smallest number (in this case, 1.64 for oxygen). This yields the molar ratios of the elements in the molecule, i.e. the chemical formula. (You may need to multiply this result by some factor to get the whole-number ratios). Thus the formula of the molecule appears to be $C_7H_6O_2$.

Note that this formula is actually an "empirical" rather than a "molecular" formula. A molecular formula tells us the actual number of atoms of each element in a single molecule of the compound. The empirical formula, on the other hand, gives us only the simplest whole-number ratios of the elements. Thus the molecular formula might be $C_7H_6O_2$, but could just as easily be $C_{14}H_{12}O_4$, $C_{21}H_{18}O_6$, etc. In order to determine the actual molecular formula, we would need to know the molecular weight of the compound as well.

Determine the "Double Bond Equivalency" (DBE) of the molecule. DBE or degree of unsaturation is the number of double bonds and/or rings in a molecule. In carrying out a calculation, consider halogens as hydrogen and ignore Oxygen and divalent Sulfur.

General formula:

$$DBE = \frac{(2 \times \#C) + 2 - (\#H - \#N)}{2}$$

$$\frac{(2 \times 7) + 2 - (6 - 0)}{2}$$

Unknown 7:   **DBE = 5**

So what does this mean? Some common functional groups with their double bond equivalents are given below:

Functional Group	# of Double Bond Equivalents
Alkane	0
Alkene	1
Alkyne	2
Cycloalkane	1
Benzene ring	4
Carbonyl group	1
Cyano group	2

For a compound with four or more DBE's, there is generally an aromatic ring (3 DBE's for the ring double bonds + one DBE for the ring). Five double bonds therefore suggests an aromatic ring with an attached alkene, cycloalkane or carbonyl group. For Unknown 7, we have already concluded that the compound is likely a carboxylic acid. The 5th DBE is therefore the carbonyl (C=O) of a carboxyl group.

Therefore the simplest, most likely, possibility for our unknown is a benzene ring with an attached carboxyl group, i.e. benzoic acid. At this stage, we could go directly to the Table of Unknowns for carboxylic acids and compare the physical data with benzoic acid. For the sake of illustration, however, and to confirm our conclusions thus far, let us proceed to the classification and derivative tests.

Step 5: Classification tests.

Classification tests are chemical reactions which are diagnostic for specific classes of compounds. Having narrowed the possibilities considerably by the solubility and fusion tests, we can use the classification tests to positively identify the class to which our unknown belongs. You have already performed a number of classification tests in the lab, including for example the Lucas test for alcohols and the 2,4-DNP test for aldehydes/ketones.

Select Classification Tests from the Experiments Menu. At this stage, the help screen is absolutely essential. Click on Help on Classification Tests from the Help menu. Based upon the tests already performed, you may at this step narrow the list of possibilities by eliminating a number of potential classes.

For example, for Unknown 7, we know that the compound is acidic. Therefore we can eliminate the neutral and basic groups, narrowing the possibilities to phenols and carboxylic acids. Although the Double Bond Equivalency and the fusion tests suggested a carboxylic acid, let's test first for a phenol. Two possibilities are listed: ferric chloride and bromine-water. You performed both tests in the lab in Expt. 5. Look back now to see the chemical reactions for these tests and the information to be gained from them. (Help screens explaining each test can be accessed after the test has been selected).

Select the Ferric Chloride Test from the Classification Tests Menu. As expected, this test gives a negative result, in keeping with our previous conclusions.

Look at the list of tests for carboxylic acids. We have already checked the solubility in sodium bicarbonate. The equivalent of the silver nitrate test has already been done, when we added $AgNO_3$ to the fusion mixture to check for halides. The "false positive" halides test was an indication of a carboxylic acid. It remains only to check the pH of an aqueous solution of the compound. *Check the pH* now. Does it help confirm or to deny our tentative identification of the unknown as a carboxylic acid?

Step 6: Derivatives.

We have now established the major functional group and a possible structure for our unknown. Based upon this information and the melting and/or boiling point and empirical formula of the molecule, we may be able to identify it at this point. Often, however, there will be two or more possibilities in the same class having very similar melting/boiling points or molecular formulas. It is often useful, then, to react the compound to form a derivative. For example, the 2,4-DNP test for aldehydes/ketones results in a solid derivative, a "dinitrophenylhydrazone." The melting point of solid derivatives of the unknown can be used to distinguish between two or more similar compounds.

Select Derivative Tests from the Experiments Menu. You decide which derivative tests to perform by selecting the help screen for the derivative tests from the help menu (the last item on the menu, as usual). For carboxylic acids this includes a p-toluidide, an anilide and an amide. Help screens are available for each derivative showing the chemistry involved. You should always provide data for at least 2 derivatives. Unfortunately, in some cases, not all derivatives are solids. If only 1 solid derivative is available, indicate the fact that the others are not solids in your lab report. The melting points and color of the derivatives are listed in your lab notebook.

We are now ready to consult the on-line files of unknowns. These are a listing, by class, of compounds with their boiling and/or melting points and the melting points of their solid derivatives.

Select Derivative Test Data from the Help Menu. Select carboxylic acids and scroll down to the approximate melting point or boiling point for your unknown. For unknown 7 (mp = 118 to 121 degrees), there are two possibilities, Phenylhydroxy acetic acid (common name mandelic acid) ($C_8H_8O_3$) and benzoic acid ($C_7H_6O_2$). Compare this to the empirical formula which was calculated earlier. Which compound best matches the empirical formula? By matching up the melting points for the unknown and the derivatives to the 2 possibilities, it is clear that the unknown is benzoic acid.

Step 7: Infrared Spectrum

In may cases it is possible to identify the unknown solely on the basis of interpretation of the IR spectrum, since we are dealing with a very limited data base. In the real world, however, this is seldom done, since IR spectra can be very similar for different materials and a true IR expert may have difficulty in making a positive identification from the IR spectrum alone. While most modern labs use several spectroscopic methods, rather than wet chemical tests, to identify an unknown material, wet chemical techniques are still very important and for us, emphasize the organic chemistry we learn in lecture. ***It is thus important that you do not fall into the trap, in the lab or when using IOC, of obtaining an IR first and using that data to select an appropriate test. This completely eliminates the learning experience the lab and IOC is ment to impart.*** Use the IR data to confirm your wet chemistry data. Remember, your lab report must contain the wet chemical data and the lab notebook in IOC will print out the order in which you performed your tests.

The IR spectrum is available from the experiments menu (we will not use the mass specta or NMR data). Pointing the mouse at a peak will tell you the frequency (wavenumber) associated with an absorption. You will type in the assignment for each important peak in the IR analysis table below the spectrum. After you have finished with the analysis of the spectrum, click on Save Your IR Analysis and your analysis and comments will be saved to the lab notebook. A help screen is also available for the IR, to help you in assigning peaks to specific functional groups. The IR will be

138

included in the lab notebook report when it is printed.

Step 8: Solution of the Unknown

When you think you have identified an unknown, you can check to see if you are correct. Select "Answer Trial" from the Experiments Menu. Type in your answer (be sure to spell the name exactly as it is spelled in the derivatives table), and click OK. If you have successfully identified the unknown, a congratulations message will be displayed. You may also see a Windows error message, if the computers you are working with does not have a sound card. Ignore this and close the message by clicking OK.

Print out the lab notebook from the File, Print Menu. Be sure to click on the Print IR Spectrum box to include the IR in the lab notebook. Turn in the lab notebook as part of your unknowns report.

Assignment 2: Interpretation of Infrared Spectra

In the laboratory, the identification of your liquid unknown should be based primarily upon the results of the wet chemical tests which you perform. We will, however, use IR spectroscopy as an example of instrumental analysis methods. This exercise will give you practice in the interpretation of IR spectra.

Attached are four IR spectra, along with their molecular formulas and some key solubility/wet chemistry data. A systematic approach to the interpretation of IR spectra is found in Section IX of this expeiment. You should also read the material on IR spectroscopy found in your lecture text.

Following the procedures in the lab manual, identify the functional groups responsible for as many of the major absorption bands as you can. Indicate, directly on the spectra, what each major absorption represents (i.e. Saturated C-H Stretch).

What absorptions are missing? What does this tell you about the structure of the molecule? Does your interpretation correlate with the molecular formula/chemistry data.

Construct a detailed, step by step proof for the structures you assign to the unknowns, using chemical, molecular formula (including double bond equivalencies) and wet chemical data. Draw the chemical structure and name the chemical. If you cannot make an absolute structure assignment because isomers are possible, draw examples of the remaining isomers.

Infrared Spectroscopy Homework Problems

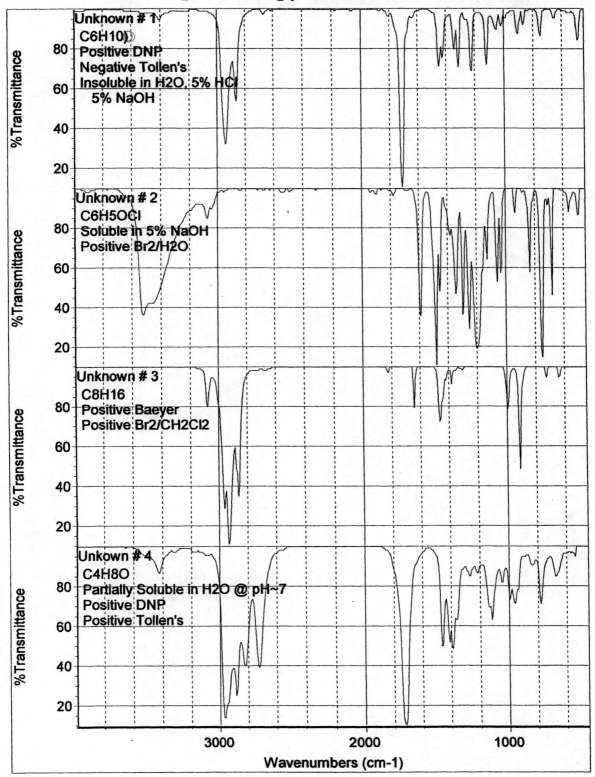

Unknown # 1
C6H10
Positive DNP
Negative Tollen's
Insoluble in H2O, 5% HCl
5% NaOH

Unknown # 2
C6H5OCl
Soluble in 5% NaOH
Positive Br2/H2O

Unknown # 3
C8H16
Positive Baeyer
Positive Br2/CH2Cl2

Unkown # 4
C4H8O
Partially Soluble in H2O @ pH~7
Positive DNP
Positive Tollen's

%Transmittance

Wavenumbers (cm-1)

141

You will be given two unknowns to analyze in the lab. The first unknown, a liquid, will be analyzed in the laboratory using a combination of physical, chemical and spectral methods. The second unknown, a solid, will be identified from a short list of possibilities by obtaining a melting point in the laboratory and comparing it to those of the known chemicals.

A. Experimental Outline

You will use the following sequence for the analysis of your unknown:

1) Obtain the melting point for your solid unknown. You may, if you have time, confirm the identity of the solid by mixing a small amount of it and the compound you believe it to be, and retaking the melting point. If the new mixed melting point does not change, the identity of the unknown has been confirmed.

   unknown #2 2- naphthal

2) You will obtain a boiling point, note the odor, color and viscosity and perform the solubility tests for the liquid unknown. These physical tests will allow you to narrow down the list of possible unknowns (there are 36 possible unknowns listed in section XII) to a small number.

   burn test - sooty flame   bp ≈ 75°C = 70°C No acid - Litmus paper   Neutral / basic - PH paper

3) Perform the wet chemical functional group tests indicated by the physical tests previously carried out.

   2,4 DNP - No aldehyde, ketone

4) Obtain an infrared spectrum for your unknown. This should be the **last** analysis you perform in the lab. If you obtain the IR before completing the wet chemical tests, ***do not be tempted to declare the identity of the unknown based solely on the IR data***. Misinterpreting the IR, which is easy to do even for an experienced chemist, may lead you to omit performing a crucial solubility or wet chemical test necessary for the identification. Compare the infrared spectrum you obtain to the list of IR spectra in section XII to find a match. Remember, you will be required to **analyze** the IR in detail for your report.

5) Prepare a report, **proving** the identities for your three unknowns.

Each of these steps is detailed below:

IV. The Report

You and your partner should turn in a report outlining **your proofs** for the identification of the three unknowns. The report should be typed, if possible, double spaced and 9 - 10 pages in length, not counting the cover page, spectra or the data sheets. The report should contain one section for each of the unknowns. Each section should include an introduction, a chemical/ physical analysis section, a spectral analysis section and a concluding section. You should attach the labeled, analyzed spectra and filled-in data sheets for each unknown after the concluding sections.

The cover page should include the course title and number, the name of the experiment, your names (with signatures) and the date of submission. The introduction should identify the unknown (Unknown 8, etc.) and generally discuss your approach to the identification of the unknown.

The chemical/physical analysis section should list each test performed, the results and their meaning.

The IR spectral data should be fully analyzed as outlined later. **Do not** expect your instructor to analyze the spectra.

The concluding section should organize the above data into a logical proof of the structure for the unknown. Remember, you will be graded for your proof, not for the actual identity of the molecule.

V. Physical Analysis

Analysis of the physical properties of a material will often give many clues to its identity. Observe the color; physical state, odor, viscosity, volatility and melting or boiling point of your unknown.

A. Physical State

Is the molecule a liquid or a solid. If the material is a liquid at room temperature, this will obviously eliminate any possible solids. Beyond this obvious fact, the physical state of the material is also related to its chemical structure and molecular weight. How? Compare the melting and boiling points of straight chain and branched chain alkanes, for instance, or the difference in boiling points of H_2O and H_2S.

B. Cautiously Smell your Unknown

Your instructor will show you the proper technique for doing this. Functional groups often give a unique odor to chemicals. the smell is often enough information to identify a material (vanillin, cinamaldehyde). In order to aid you in correlating functional groups with odor, examine the reference chemicals set out for you in the laboratory.

C. Observe the Color of the Unknown

Most pure organic chemicals are colorless, most solids clear or white. Some chemicals are yellow, such as nitroaromatics, due to extended conjugation systems. Some chemicals, such as the amines, react with oxygen to form highly colored impurities. Again, correlate the color with the functional groups for the reference samples in the lab.

144

D. Observe the Viscosity and Volatility

Viscosity and volalility will often correlate to the MP and BP. Low viscosity and highly volatile methylene chloride has a low B.P. The high viscosity and low volalility of mineral oil (or mixture of alkanes) indicates a high B.P.

E. Melting Point and Boiling Point

Since these values are very accurately known for pure chemicals, their determination will often limit the unknown to a very short list.

1. Melting Point

You will obtain the MP for your solid unknown, and compare its value to the MP values tabulated at the end of this experiment. You will determine the MP using a Mel-Temp apparatus by placing a very small amount (2mm) of powdered solid. One word of warning! We use the term **melting-point** and literature references such as the CRC Handbook often list a single temperature for the MP. Actually this is the midpoint of an actual melting point range. You will **report** the **MP range**. The MP range is the temperature at which you just observe melting to the temperature at which melting is complete. For a very pure chemical the MP range will usually be sharp (a very narrow range). Impure chemicals will have a broad MP range, usually lower in temperature than the literature value.

Take two melting points. The first will be done rapidly (20° per min) and will not be very accurate, since the solid and the thermometer are not in equilibrium. Lower the thermometer to just below the temperature at which you first observed melting and then adjust the Mel-Temp to give a temperature increase of 1-2°/min. **Use a new solid sample.**

When you are finished, place the used MP tubes in the designated container and **then and only then** compare your MP to the one listed in section XI. If you have time and if you feel confident that your MP indicates that you have one of the listed chemicals, mix an equal amount of your unknown and the pure compound you believe your unknown to be (a small spatula tip of each, in a reaction tube) and retake the melting point. If the 2 chemicals are identical, the MP should be the same as you originally obtained. If the MP is lower (usually 5 or more degrees) and the MP range large (usually 5-10 degrees), this indicates the 2 chemicals are different, and you need to try a mixed MP with another known. Consult you instructor first, however.

2. Boiling Point

You will use the same distillation set-up you used for the cyclohexene experiment to determine the BP of your liquid unknown. Remember to accurately place the thermometer so that the **top** of the bulb is even with the **bottom** of the side arm. Distill the liquid only until you obtain a consistant BP You should distill a few drops of the liquid into the receiver flask. Remember, the BP is affected by the atmospheric pressure. The lower the pressure, the lower the BP.

Please note! Many chemicals, including aldehydes, ketones, ethers and hydrocarbons, react with air to form peroxides, which explode when heated. To avoid this, do not distill the liquid to dryness. Only distill until the temperature stabilizes. Stop before the flask is empty.

VI. Chemical Analysis

The chemical analysis will consist of the tests for functional groups you have performed during the term. Some functional groups, such as the esters, we have not tested for and will be determined by elimination and through IR analysis.

A. Solubility tests

Solubility tests are carried out as in previous experiments, by adding 5 drops of the unknown to 0.5 mL solvent in a test tube. Use the following separation diagram to categorize the functional groups in your unknown.

Perform the solubility tests in the numbered sequence, **only until the compound dissolves** (or reacts with the reagent and dissolves).

1. Group I, Soluble in Water: Use the *blue dyed water* in order too clearly determine whether your unknown is soluble or insoluble in water. This includes small molecular weight and/or highly polar molecules such as CH_3OH or CH_3CO_2H. Check the water layer with pH (not litmus) paper to see if the material is an acid (RCOOH or Ar-OH) or base ($R-NH_2$).

2. Group II, Insoluble in Water, Soluble in 5% NaOH. The unknown is either a phenol or a carboxylic acid. Test for solubility in 5% $NaHCO_3$. If it is soluble and CO_2 bubbles are observed, the unknown is probably a carboxylic acid (RCOOH)

3. Group III, Insoluble in 5% $NaHCO_3$ but soluble in 5% NaOH. The unknown is probably a phenol (Ar-OH)

4. Group IV, Insoluble in 5% $NaHCO_3$ but soluble in 5% HCl. The unknown is probably an amine ($R-NH_2$).

146

5. Group V, Insoluble in 5% HCl but soluble (or reacts and becomes soluble) in cold, concentrated H_2SO_4. The unknown contains O, N, S or a double bond (alkenes, alkynes, alcohols, ethers, aldehydes, ketones, esters, amides, nitriles). The solution may turn a dark color.

6. Group VI, Insoluble in H_2SO_4. The unknown is probably an alkane, an aromatic or a halogen substituted alkane or aromatic.

Dispose of all hazardous waste in the non-halogenated hydrocarbons waste container, unless instructed otherwise.

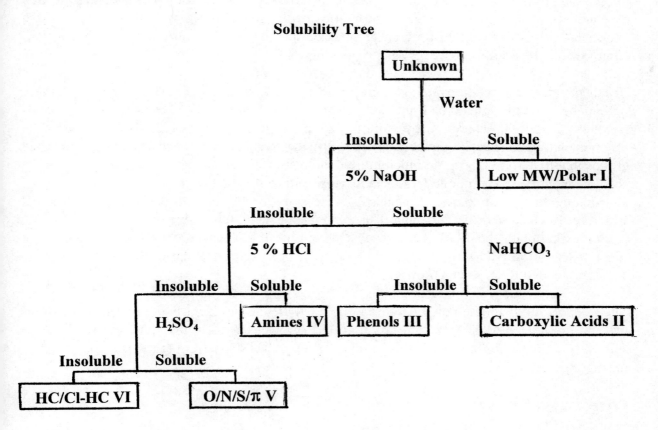

Solubility Tree

B. Functional Group Chemical Tests

The physical appearance, odor, boiling point and especially the solubility tests should enable you to decide which chemical tests are or are not suitable for your unknown. For instance, if the material was shown to belong to Group V, you might choose to perform the Baeyer, tollens, DNP, Lucas and Chromic acid tests but not the $FeCl_3$, $NaHCO_3$ or benzoyl chloride tests. Refer to the appropriate chapter for the chemical tests. Remember, *we have not performed tests for esters, amides, nitro compounds or nitriles*. You would identify these functional groups by a process of elimination and by using infrared spectroscopy.

VII. Infrared Spectroscopy

In industry, infrared spectroscopy is usually the first tool used to identify or confirm the identity of most organic and many inorganic compounds. In dealing with pure materials, such as the unknowns in this experiment, it may be the only tool necessary for absolute structure determination. Then why do we still use the wet chemistry techniques in the lab? To emphasize, for the student, the chemistry of the functional groups covered in the lecture portion of the course. In other words, to help reinforce the concepts learned in lecture. It must also be remembered that the physical properties and uses of all organic materials; plastics, elastomers, fuels, lubricants, etc., are determined by the chemical properties of the functional groups contained in these materials. This is why the IOC Program emphasizes the functional group approach, so that its use compliments the traditional wet chemistry techniques.

The theory and interpretation methodology for IR spectroscopy are covered in the lecture, so these will not be covered again here. It should be kept in mind, however, that the IR absorption bands result from IR light absorption by vibrating atoms, especially the atoms associated with the functional groups. Thus, IR can be used to confirm the presence or absence of specific functional groups in a material. IR can do much more than that, however. It can be used to establish the absolute identity of a compound, or even mixtures, since the total IR spectrum is dependent on, and unique for a particular chemical substance since each material has a unique chemical structure. IR can also be used to quantify a component in a mixture, using the computer as an aid. In this experiment, the IR will be used to determine the presence or absence of the major functional groups and, with the aid of a library of IR spectra and the wet chemistry data, to confirm the identity of the unknown.

The IR spectrum that is run on the liquid sample will be used to identify the **functional groups** that are present. You should carefully correlate the wet chemical test data, the IR data, the solubility data and boiling point before you compare your IR spectrum to the spectra at the end of this experiment (Section XII).

A. Preparing the Sample

Since only liquid unknowns will be analyzed in the lab, a simple liquid film will be used for the sample. The sample is prepared for analysis by placing *1 drop* of the unknown from a Pasture pipette onto a sodium chloride window. **(CAUTION !! DO NOT TOUCH THE FLAT SURFACES OF THE CRYSTALS-HOLD THEM BY THE EDGES)** a second window is placed over the first and the pair placed into a holder which also gently clamps the windows together. The sample prepared in this fashion, commonly referred to as a neat sample, is placed in the sample beam and a transmission spectrum obtained. This means that the IR light is transmitted through the pure (neat) sample and light absorbed at those wavelengths, or frequencies, which are identical to the vibrational frequencies of the bonds of the functional groups. You should use the mouse to click on the sample button on the computer monitor. After the background has been obtained (approximately 5 seconds), place your sample holder in the holder slots in the compartment, click yes to collect the sample and acquire your spectrum. After obtaining the spectrum, click on the print button. While the spectrum is printing, have your partner clean the crystals. **The crystals dissolve in water, do not wash with water. Please wash the crystals with isopropyl alcohol, holding the crystals with the special pair of tongs provided.** Do not drop the crystals. The crystals are placed in a desiccator, if you are the last one to use them. Pat the crystals on clean paper towel and the solvent will evaporate enough for the next student to use the crystals.

VIII. Operating Instructions For The FTIR

A. Operation of the FTIR

1. Remove the top compartment access cover on the spectrometer, making sure there is nothing in the sample holder. Replace the cover immediately and click on **Col Sam Button.** The spectrometer will scan 8 times to obtain a background spectrum. Note the absorptions for water, CO_2 and the thin organic protective coating on the KBr spectrometer compartment windows. Also note that the spectrum is curved, since the source (a glowing wire) does not put out a constant amount of light energy throughout the spectrum. The greatest amount of energy is emitted with this instrument at about $1200\,cm^{-1}$. A Window will then appear asking you to prepare to run a sample.

2. Place your sample into the sample holder through the top compartment access cover. Immediately replace the cover and use the mouse to click on OK Another 8 scans will be acquired, the co-added background spectra will be subtracted (by the computer) from the co-added sample spectra and a true IR will be displayed. This takes about 5 seconds. Click on the **PRINT** button and in about 30 seconds you will have your hard copy. One partner should clean the crystals during the printout and tell another group that the instrument is ready to use. **If necessary, show the next group how to use the instrument. REPLACE THE SAMPLE COMPARTMENT COVER** between runs. **Complete a manual interpretation of your spectrum and then find the match-up for your spectrum with the spectra at the end of this experiment**

IX. Analysis of the IR Spectrum

Three approaches are used in the analysis of IR spectra. These are:

Pattern Recognition
Positive Interpretation
Negative Interpretation

A discussion of each of these approaches follows:

A. Pattern Recognition

As the name implies, pattern recognition is the process of analyzing the spectrum as one entity. Thus, it is the entire spectrum that one is comparing to a reference, whether in a library or in the mind. Frankly, this may not really happen for the vast majority of the spectra that are analyzed, especially by "experts". Comparison of a spectrum with a literature reference, with no additional analysis, probably only occurs in quality control labs, where individuals with little actual training in IR are required to determine whether incoming shipments of plastics, solvents or other raw materials meet purity standards. Most QC labs do not have the responsibility of determining "what" is wrong with a shipment. That is the duty of the research or analysis groups. Determination of the acceptability of a material may be afforded by comparison of its IR spectrum with that of a standard. If the IR spectrum of the material in question and the standard are obtained under the same conditions and with the same technique, they should be identical. The presence of contaminants, the absence of an additive or the wrong ratios of all ingredients will be made obvious by a difference in the spectra or by computer spectral substraction. However, this really doesn't involve spectral interpretation to determine the actual composition of the material.

Then what about the expert, who can identify spectra by simply looking at it for a few seconds? I don't think that the process of identifying the sample is quite so simple. In fact, the identification process for the expert is probably a complicated process of the use of positive and negative interpretation to determine the family of chemical that the material belongs to (more than likely he is also using every other piece of data available for the material as well, i.e. the history of the sample) and then perhaps sorting through the stack of spectra neatly filed away in his memory. Obviously, the expert would not be able to deduce the exact structure for a sample he had not seen before, without first analyzing the spectrum in painstaking detail, and probably using corroborating analytical techniques to back up his analysis.

Thus, while you may not have the years of experience, and perhaps, the computer-like memory of the expert, with the proper use of positive and negative identification, and a good library of spectra, with a little practice you should be able to easily determine: (1) the family the sample belongs to and (2) the actual structure or one very close to it.

B. Positive Identification

Each organic (and inorganic) functional group has a specific set of stretching and deformation vibrational modes for which there are absorption frequencies in the IR. Many of these functional group vibrations occur at approximately the same frequencies, no matter what the actual structure of the molecule might be. Thus, the presence of an absorption at a particular position, intensity and with a particular shape, is indicative of the presence of the functional group in the sample. This, in effect, is positive identification. For instance, the carbonyl group (C=O) is usually the strongest absorbing band in the IR spectrum, occurring at approximately 1725 cm^{-1}. If a strong band is observed in the IR at 1740 cm^{-1}, we can say that the sample contains a C=O group.

C. Negative Interpretation

Negative interpretation, as the name implies, involves noting the "absence" of an absorption band and therefore, interpreting this to mean that the functional group which would have been responsible for the band would not be present in the material in question. Thus, the absence of absorption in the region from 3000-2800 cm^{-1} would mean that the molecule would not have aliphatic (saturated) C-H.

A word of warning - remember that the absorption of IR energy by a functional group is dependent on the dipole moment change which occurs during a vibration. If a vibration is totally symmetrical, there will be no IR absorption, even though the functional group is present in the molecule. The dipole change, for many more functional groups, is small. As a result, many groups have very weak absorptions, which might be missed in the noise or covered up by other absorptions in the molecule. This problem can be checked by the use of Raman spectroscopy, in which small or no dipole changes (symmetrical bonds) result in strong absorptions. Unfortunately, very few labs have access to Raman equipment.

A second word of warning concerning materials which are mixtures: functional groups, which do not strongly absorb for one component, may be totally hidden by the stronger absorptions of a second component. In some cases, a component may account for 30% of the mixutre, but it's IR spectrum might be totally obscured by a strongly absorbing component. There is also the possibility that chemical interactions (H-Bonding, etc.) Will cause large shifts in spectral bands.

D. Combined Interpretation

With practice, a spectroscopist will not often have to think too deeply about the process of interpretation. The actual interpretation process, subconscious or not will actually involve a combination of positive and negative interpretation, along with the utilization of the background data for the sample, to arrive at a conclusion as to the family to which the sample belongs. At that point, the spectroscopist then uses pattern recognition (of memorized spectra), a library comparison, or, if no library fit is found, a detailed analysis of as much of the remaining spectra as possible to determine the actual structure of the sample.

E. The Fingerprint Region

Unfortunately, not all absorption bands are so constant in their position, intensity or shape. This is especially true for deformation vibrations. The vibrating atoms often cause vibration of adjacent atoms (addition and subtraction bands), which causes the vibrational frequencies to depend on the actual structure of the molecule in question. Thus, very few of the predicted fundamental vibrations in a molecule can be used for positive identification of a functional group in a molecule. However, enough vibrations are consistent in their position, intensity and shape to allow the identification of the family that the chemical belongs to. The remaining many absorptions which are not consistent usually fall in the range from 1500-250 cm^{-1}, which is called the *Fingerprint* region.

Comparison of the entire spectrum, especially the fingerprint region, with that of the reference spectrum, obtained with the same sampling technique, enables a positive (or negative) confirmation of the structure or component makeup of the sample.

One word of warning - it is entirely possible that within the resolution of the IR instrument utilized for either the sample or reference spectra acquisition, two totally unrelated molecules, which have the same functional groups but differing structures, might have spectra which appear to be the same. Of course, there would have to be differences, but they might be small enough that they would be overlooked. This is why it is of the utmost importance to consider the history, physical and chemical properties of the material in question.

F. Key Absorption Band Analysis

At the point where you are ready to examine the spectrum and perform a family analysis, it is presumed that you have already accumulated the necessary information on sample history and chemical and physical data. This will enable you to determine whether your analysis is leading along the path toward a correct structure analysis.

We now need to examine the IR spectrum in a systematic fashion. The actual system will, to some extent, be determined by the types of samples you are working with. For example, if you are working exclusively with perfluorinated hydrocarbons, you will probably want to direct your attention toward the C-F vibrations in the spectrum. If you are dealing with inorganics, you may need to focus on the far IR (250-50cm^{-1}). However, the following list of functional groups should always be considered when analyzing a spectrum, even if just for a negative analysis, in order to ensure that the major families of organic materials are covered.

The list should be covered in the order presented. Remember, each band falls within a fairly consistent range. However, each functional group can be subcategorized, leading to more specific ranges over which to look. Some of these subcategories, such as that for the carbonyl (C=O), can be found in texts dealing with IR.

The following bands are all stretching modes. There are many more deformation vibrations for each functional group, some of which are quite characteristic and useful. However, the stretching vibrations are much more reliable, as a class.

The given band position is an average value for the functional group, which can vary over the given range, depending on the physical-chemical environment of the group and sometimes the sampling technique employed. The value inside the perenthesis indicates the intensity of the absorption: s = strong, m = medium, w = weak.

Finally, remember that the band factors that are important in analysis of a spectrum are:

Absorption Band Position (average value or range in cm^{-1} or μm)

Absorption Band Intensity (ie strong[s], medium[m], or weak[w])

Absorption Band Shape (sharp, broad or very broad)

Correlation by Band Position

FUNCT. GP	BAND POSITION	COMMENTS
1.) C=O	1725 cm^{-1}	Range: 1900-1600 cm^{-1}, usually the strongest band, usually moderately broad.
2.) Sat. C-H	2950 cm^{-1}	Range: 3000-2800 cm^{-1}, usually several overlapping bands.
3.) Unsat. C-H	3050 cm^{-1}	Range: 3100-3000 cm^{-1}, both Ar-H and C=C-H, (m-s).
4.) C=C	1650 cm^{-1}	Range: 1680-1620 cm^{-1}, sharp, (m).
5.) C=C Aromatic	1600 & 1500 cm^{-1}	Also possibly 1580 & 1450 cm^{-1}, sharp, (m).
6.) O-H	3500 cm^{-1}	Range: 3650-3200 cm^{-1}, alcohols & phenols, sharp (no H-bonding, dilute soln.), broad (H-bnding, conc. Soln., reduced by steric hindrance), (s).
7.)N-H	3400 cm^{-1}	Range: 3500-3300 cm^{-1}, primary amine at 3500 & 3400 cm^{-1}, secondary-one band, also amides

8.) Alkyne C-H	3300 cm^{-1}	Terminal alkyne, sharp, (s).
9.) Aldehyde C-H	2850 & 2750 cm^{-1}	Range: 2900-2700 cm^{-1}, often hidden under aliphatic C-H, sharp, (w).
10.) COO-H	3000 cm^{-1}	Range: 3400-2500 cm^{-1}, very broad, (s), often covers all other bands in the region, disappears upon conversion to a salt.
11.) Alkyne, Cyano, 2250 cm^{-1} Isocyanate	2250 cm^{-1}	Range: 2260-2100 cm^{-1}, sharp, (w-m).
12.) Si-H	2160 cm^{-1}	Range: 2250-2100 cm^{-1}, higher frequency if electronegative Atom attached to Si, sharp (m-s).
13.) Si-Me	1260 cm^{-1}	Range: 1270-1250 cm^{-1}, sharp, (s).
14.) Si-O-Si	1020 & 1090 cm^{-1}	Very prominent, (s).
15.) C-O	1100 cm^{-1}	Range: 1275-1000 cm^{-1}, broad, prominant, two bands for R-O-R, often difficult to assign since in the fingerprint, broad, (s).
16.) C-Cl	700 cm^{-1}	Range: 800-600 cm^{-1}, broad, (s).
17.) Fingerprint	1500-250 cm^{-1}	Positive confirmation of the structure by comparison between the sample and the reference, with identical sampling techniques.

There are also three deformation (bending vibration) bands which are fairly reliable, although they are in the fingerprint region and thus should be used for confirmation of information in other bands (i.e. saturated C-H stretch)

FUNCT. GP	BAND POSITION	COMMENTS
18.) CH$_2$	1460 cm^{-1}	Range: 1485-1445 cm^{-1}, scissoring vibration, overlaps with CH$_3$ asymmetric bend, sharp, (m).

19.) CH_3	1460 cm^{-1}	Range: 1470-1430 cm^{-1}, asymmetric bend, overlaps with CH_2 scissoring, sharp, (m).
20.) CH_3	1375 cm^{-1}	Range: 1380-1340 cm^{-1}, symmetric bend, sharp, (m).

G. Correlation by Spectral Region

Some spectroscopists prefer a second approach, in which the spectrum is analyzed by region. A close examination of this approach and the previous one indicates that, in reality, they are the same, with the exception that the band position technique looks at the regions in a different sequence.

WAVENUMBER (cm^{-1})	FUNCTIONAL GROUPS
1.) 3700-3100	O-H, N-H, Alkyne C-H
2.) 3100-3000	C=C-H, Ar-H, cyclopropyl C-H
3.) 3000-2700	Saturated C-H, Aldehyde C-H
4.) 2300-1900	C=C, C≡N, N=C=O, Si-H
5.) 1850-1600	C=O
6.) 1700-1550	C=C, C=N
7.) 1600 & 1500	C-C Aromatic
8.) 1460	CH_2, CH_3 assymetric bend
9.) 1375	CH_3 symmetrical bend
10.) 1300-1000	C-O, Si-O, Si-CH_3
11.) 1000-600	unsaturated, Aromatic C-H wag, C-Cl

X. Example IR Spectra

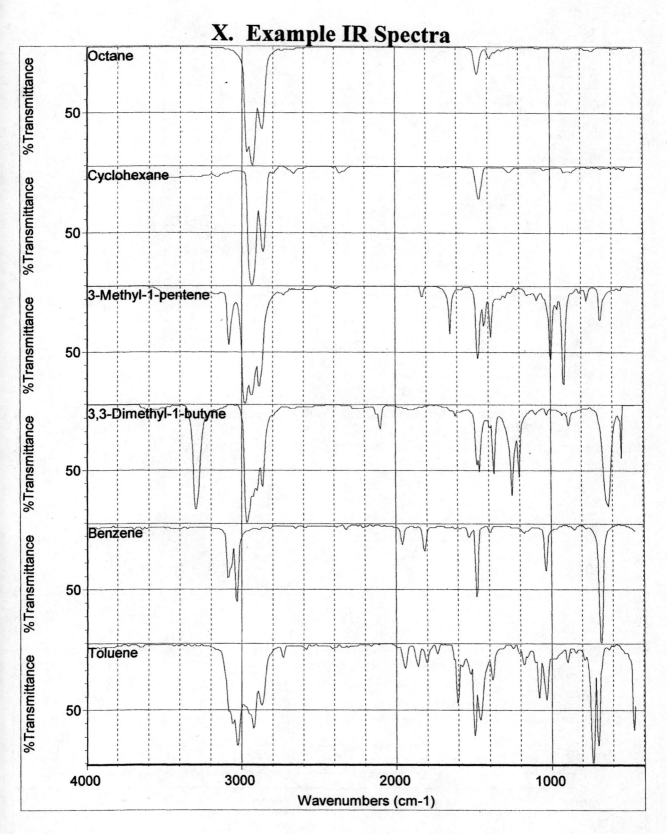

Octane

Cyclohexane

3-Methyl-1-pentene

3,3-Dimethyl-1-butyne

Benzene

Toluene

%Transmittance

50

Wavenumbers (cm-1)

4000   3000   2000   1000

157

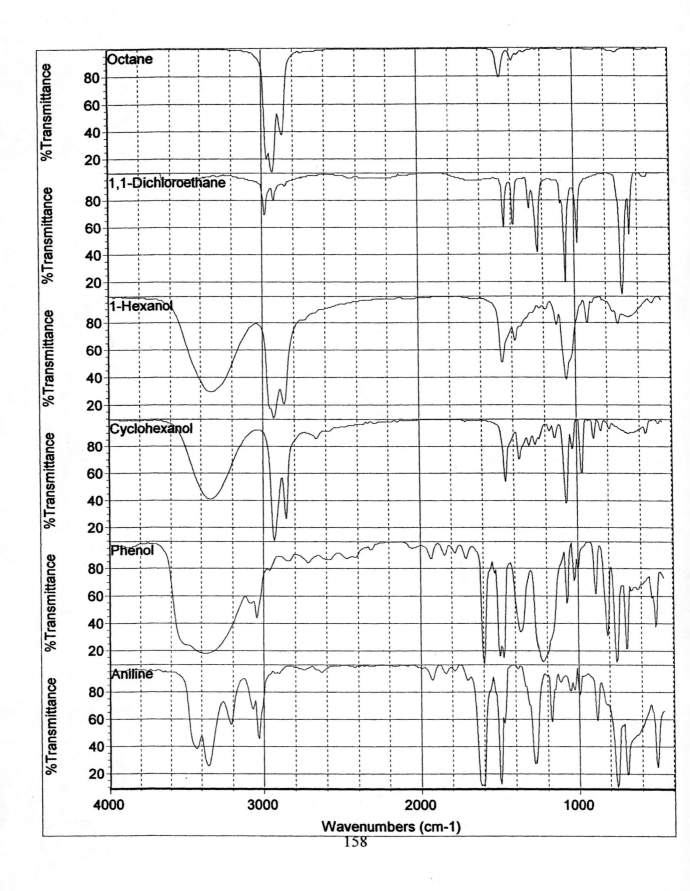

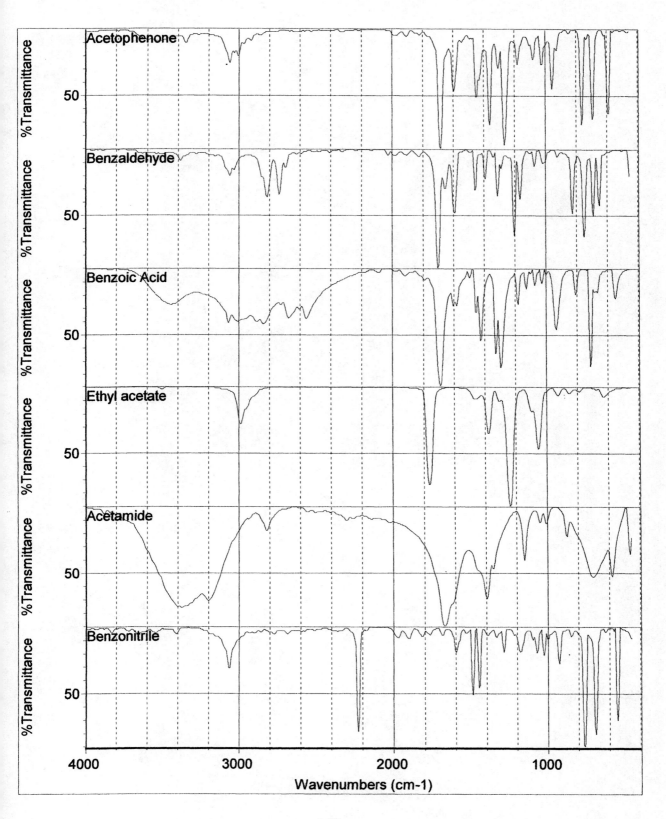

XI. Liquid and Solid Unknown Lists

A. Possible Liquid Unknowns for CHEM-146

	Compound	Alternate Name	BP
1	tetrahydrofuran	THF	65
2	butyraldehyde	butanal	76
3	ethyl acetate	ethanoic acid ethyl ester	76
4	butylamine	1-aminobutane	78
5	methyl ethyl ketone	2-butanone	80
6	cyclohexane		81
7	cyclohexene		83
8	isooctane	2,2,4-trimethylpentane	99
9	2-butanol	sec-butyl alcohol	100
10	formic acid	methanoic acid	100
11	methylcyclohexane		101
12	3-pentanone	diethyl ketone	102
13	valeraldehyde	pentanal	103
14	toluene	methyl benzene	110
15	4-methyl-2-pentanone	methyl isobutyl ketone	117
16	1-butanol	n-butyl alcohol	117
17	acetic acid	ethanoic acid	118
18	1-octene		122
19	butyl acetate	ethanoic acid butyl ester	126
20	hexylamine	1-aminohexane	130
21	ethylbenzene		136
22	o-xylene	1,2 dimethyl benzene	143
23	2-ethyl-1-butanol		146
24	amyl acetate	ethanoic acid 1-pentyl ester	149
25	cumene	isopropyl benzene	152
26	dimethyl formamide	methanoic acid N,N dimethyl amide	153
27	anisole	methyl phenyl ether	155
28	butyric acid	butanoic acid	162
29	2,6-dimethyl-4-heptanone	diisobutyl ketone	168
30	benzaldehyde		178
31	limonene	1-methyl-4-isopropenyl cyclohexene	178
32	benzylamine		185
33	valeric acid	pentanoic acid	186
34	o-cresol	2-methyl phenol	191
35	1-octanol	n-octyl alcohol	194
36	p-cresol	4-methyl phenol	202

B. Possible Organic Unknown Solids

Solid Chemical	MP (°C)
lauric acid	44-46
chloroacetic acid	61-63
acetamide	79-81
azelaic acid	109-111
2-naphthol	122-123
malonic acid	135-137
cholesterol	148-150
salicylic acid	158-161
d-tartaric acid	172-174
succinic acid	188-190
phthalic acid	205 dec
caffeine	234-236

1.

XII. Liquid Unknown IR Spectra

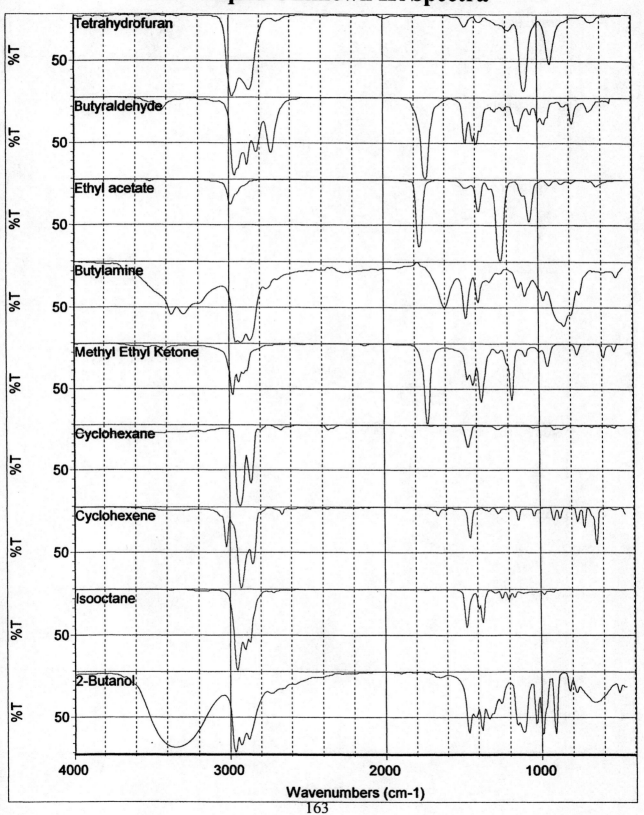

Wavenumbers (cm-1)

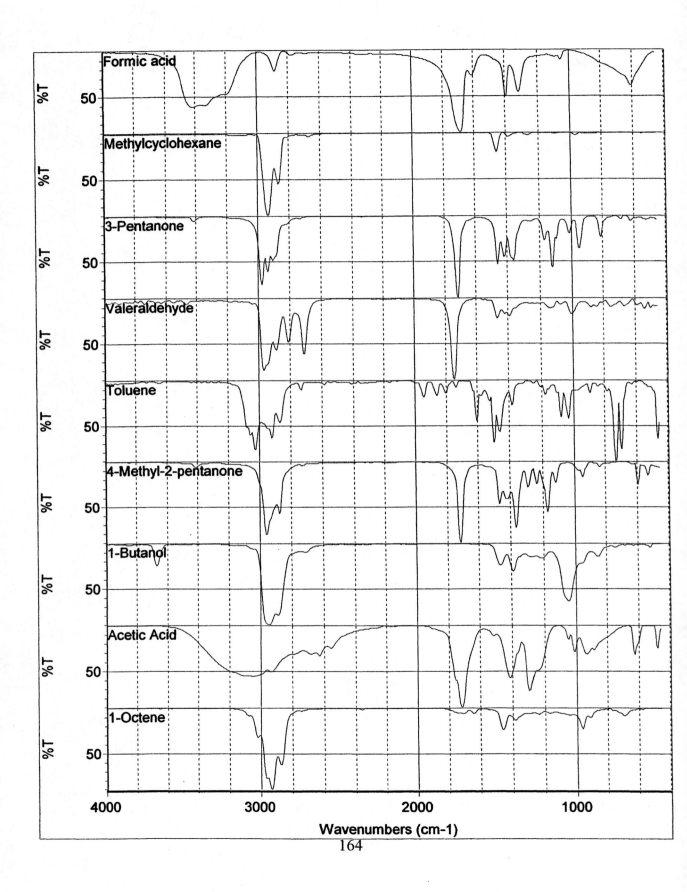

Formic acid

Methylcyclohexane

3-Pentanone

Valeraldehyde

Toluene

4-Methyl-2-pentanone

1-Butanol

Acetic Acid

1-Octene

%T 50

Wavenumbers (cm-1)

4000 3000 2000 1000

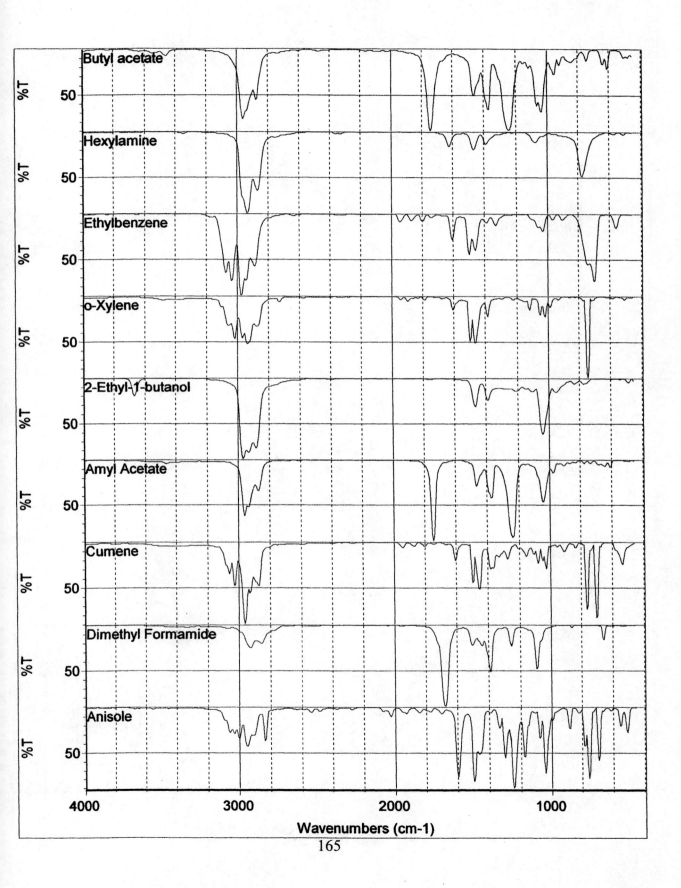

Butyl acetate

Hexylamine

Ethylbenzene

o-Xylene

2-Ethyl-1-butanol

Amyl Acetate

Cumene

Dimethyl Formamide

Anisole

%T — 50 (repeated for each spectrum)

4000   3000   2000   1000

Wavenumbers (cm-1)

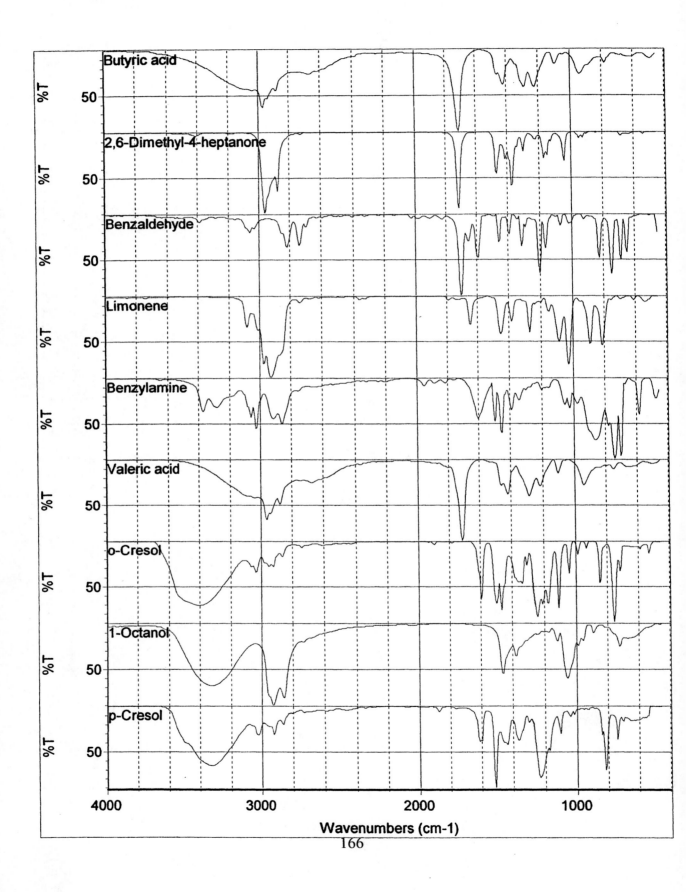

Butyric acid

2,6-Dimethyl-4-heptanone

Benzaldehyde

Limonene

Benzylamine

Valeric acid

o-Cresol

1-Octanol

p-Cresol

Wavenumbers (cm-1)

XIII. Unknown Data Sheets

A. Liquid Unknown Data Sheet

1. Observed Physical Properties
   a) color _____
   b) viscosity _____
   c) odor _____

2. Boiling Point
   BP = _____

3. Solubility
   a) water _____
      If water soluble, pH of solution _____
   b) If water insoluble, solubility in 5% HCl _____
   c) If insoluble, the HCl, solubility in 5% NaOH _____
   d) If soluble in NaOH, solubility in $NaHCO_3$ solution _____
   e) If insoluble in all the above, solubility in H_2SO_4 _____

4. Using the solubility tree, what functional group family does your unknown belong to?

   _____

5. On the basis of the observed physical properties and solubility data, what are the results of the **appropriate** chemical tests? Fill in those you performed.

   a) Br_2/CH_2Cl_2 test for alkenes/alkynes _____
   b) $KMnO_4$ test for alkenes/alkynes _____
   c) Lucas Test ($ZnCl_2$) for alcohols _____
   d) Chromic Acid test for alcohols _____
   e) $FeCl_3$ test for phenols _____
   f) DNP test for aldehydes and ketones _____
   g) Tollen's test for aldehydes _____
   h) Benzoyl chloride test for amines _____
   i) Flame test for unsaturation _____
   k) Flame test for halogens _____

167

6. Infrared Spectroscopy Data

   a) List the relevant observed functional group absorption frequencies (cm^{-1}), their observed shape (broad, etc.) and intensities (s,m,w,etc.) and assigned functional group. (i.e. 2250-cm^{-1}, broad doublet, w, CO_2)

Wave number (cm^{-1})	Shape.Intensity	Functional Group
_____	_____	_____
_____	_____	_____
_____	_____	_____
_____	_____	_____
_____	_____	_____
_____	_____	_____

   b) List the functional group absorptions **absent** from your spectrum.

   c) Attach your IR spectrum, labeling each assigned peak, and also attach a copy of the reference spectrum obtained from your library search.

7. Using all of your collected data, BP, odor, solubilities, etc., write a **proof** leading to the identification of the structure for your unknown, the list of possible unknowns given in the following table.

B. IOC Unknown Data sheet (Turn in a data sheet for each IOC unknown)

IOC Unknown # _____
Boiling Point/MP _____°C

1. Solubility Results:

2. Fusion Results:

3. Functional Group Identification Tests:

4. Derivative Test Results:

5. Molecular Formula and Double Bond Equivalence (show calculations):

6. IR Results

a) Functional Groups Identified

Wave number (cm^{-1})	Shape.Intensity	Functional Group
_____	_____	_____
_____	_____	_____
_____	_____	_____
_____	_____	_____
_____	_____	_____
_____	_____	_____

7. List the functional group absorptions **absent** from your IR spectrum.

8. Attach your IR spectrum, labeling each assigned peak.

9. Using all of your collected data, BP, odor, solubilities, etc., write a **proof** leading to the identification of the structure for your unknown, the list of possible unknowns given in section XII..

C. **Solid Unknown**

1. Solid Unknown # _____

2. Experimental Melting Point _____°C

3. Name and Reported Melting point of Solid_____

170